Engineering Design Methods

SECOND EDITION

Engineering Design Methods

SECOND EDITION
Strategies for Product Design

Nigel Cross
The Open University, Milton Keynes, UK

JOHN WILEY & SONS
Chichester · New York · Brisbane · Toronto · Singapore

Copyright © 1989, 1994 by John Wiley & Sons Ltd,
            Baffins Lane, Chichester,
            West Sussex PO19 1UD, England
            (+44) (243) 779777

Reprinted July 1994
Reprinted November 1994
Reprinted February 1995

Other Wiley Editorial Offices

John Wiley & Sons, Inc., 605 Third Avenue,
New York, NY 10158-0012, USA

Jacaranda Wiley Ltd, 33 Park Road, Milton,
Queensland 4064, Australia

John Wiley & Sons (Canada) Ltd, 22 Worcester Road,
Rexdale, Ontario M9W 1L1, Canada

John Wiley & Sons (SEA) Pte Ltd, 37 Jalan Pemimpin #05-04,
Block B, Union Industrial Building, Singapore 2057

Library of Congress Cataloging-in-Publication Data

Cross. Nigel, 1942–
    Engineering design methods : strategies for product
design / N. Cross.—2nd ed.
        p.   cm.
    Includes bibliographical references and index.
    ISBN 0 471 94228 6
    1. Engineering design.   I. Title.
TA174.C76   1994
620′.0042—dc20   93-30028
                                                          CIP

British Library Cataloguing in Publication Data

A catalogue record for this book is available from the
British Library

ISBN 0 471 94228 6

Typeset in 10/13 pt Palatino by Acorn Bookwork, Salisbury
Printed and bound in Great Britain by Redwood Books, Trowbridge, Wilts

Contents

Acknowledgements

The author and publisher gratefully acknowledge the following for permission to reproduce figures:

Figure 1: T.A. Thomas, *Technical Illustration*, McGraw-Hill. Figure 2: J. Fenton, *Vehicle Body Layout and Analysis*, Mechanical Engineering Publications. Figure 3: C. Moore and Van Nostrand Reinhold. Figure 4: A. Howarth. Figure 8: M.J. French, *Conceptual Design for Engineers*, The Design Council. Figures 9, 10, 11: B. Hawkes and R. Abinett, *The Engineering Design Process*, Longman. Figures 14, 23, 32, 33, 37, 40, 41, 50, 61, 62, 63: G. Pahl and W. Beitz, *Engineering Design*, The Design Council/Springer-Verlag. Figures 15, 16: VDI-Verlag. Figure 17: L. March, *The Architecture of Form*, Cambridge University Press. Figure 21: J.C. Jones, *Design Methods*, John Wiley/David Fulton. Figure 24: E. Tjalve, *A Short Course in Industrial Design*, Butterworth. Figure 25: S. Pugh, *Total Design*, Addison Wesley. Figures 30, 31: E. Krick, *An Introduction to Engineering*, John Wiley. Figures 34, 35, 36: V. Hubka, M.M. Andreasen and W.E. Eder, *Practical Studies in Systematic Design*, Butterworth. Figure 42: S. Love, *Planning and Creating Successful Engineered Designs*, Advanced Professional Development. Figures 43, 66: U. Pighini, *Design Studies*, Butterworth. Figure 44, 67: D.G. Ullman, *The Mechanical Design Process*, McGraw-Hill. Figures 45, 46: R. Ramaswamy and K. Ulrich, *Design Theory and Methodology*, ASME. Figures 47, 48: J.R. Hauser and D. Clausing, *Harvard Business Review*. Figure 49: K.W. Norris, *Conference on Design Methods*, Pergamon. Figures 51, 52: V. Hubka, *Principles of Engineering Design*, Butterworth. Figures 53, 54: K. Ehrlenspiel, ICED 87. Figures 55, 56: M. Tovey/S. Woodward, *Design Studies*, Butterworth. Figures 64, 65: M. Shahin, *Design Studies*, Butterworth. Figures 68, 69: T.C. Fowler, *Value Analysis in Design*, Van Nostrand Reinhold. Figures 70, 71: Engineering Industry Training Board. Figure 72: A.H. Redford, *Design Studies*, Butterworth. Figures 73, 74, 75: The Open University.

Introduction

This book offers a strategic approach to designing products. It is intended primarily for use by teachers and students of engineering design and industrial design. Its main emphasis is on the design of products that have an engineering content, although most of the principles and approaches that it teaches are relevant to the design of all kinds of products. It is essentially concerned with problem formulation, conceptual and embodiment design, rather than the detail design which is the concern of most engineering texts. The book can most effectively be used in conjunction with projects and exercises that require the analysis and clarification of design problems and the generation and evaluation of design solutions.

This second edition of the book has been fully revised and updated. A complete new chapter has been added on design for customer satisfaction, based around the quality function deployment method (QFD)—Chapter 7 'Determining Characteristics'. Many new examples of the application of design methods in practice have also been introduced.

The contents of the book divide into three parts. Chapters 1 to 3 offer an overview of the nature of design activity, the design process and design methods in general. Chapter 1 introduces students to the kinds of activities that designers normally undertake, and discusses the particular nature of design problems and the abilities that designers call upon in tackling them. Chapter 2 reviews several of the models of the design process which have been developed in order to help designers structure their approach to designing, suggests a new hybrid procedural/structural model, and outlines the general need for new procedures in design. Chapter 3 reviews the field of design methods, describes a number of methods that help to stimulate creative design thinking and introduces the rational methods which are presented in the next part of the book.

Chapters 4–10 constitute a manual of design methods (the

tactics of design), presented in an independent-learning format, i.e. students can be expected to learn the methods directly from the book. The seven chapters follow a strategic procedural sequence for the design process, providing instruction in the use of appropriate methods within this strategy. Each chapter presents a separate method, in a standard format of a step-by-step Procedure, a Summary of the steps and a set of practical Examples concluding with a fully worked example. The seven methods included are:

Objectives Tree
Function Analysis
Performance Specification
Quality Function Deployment
Morphological Chart
Weighted Objectives
Value Engineering

Finally, Chapter 11 outlines strategic approaches to the design process, utilizing the most appropriate combination of creative and rational methods to suit the designer and the design project. Reflecting the approach that is implicit throughout the book, the emphasis is on a flexible design response to problems and on ensuring a successful outcome in terms of good product design.

Although intended primarily for students, the book may also be useful as an introduction to design for the many teachers and practitioners in engineering who found this subject sadly lacking from their own education, and as a general guide to the field of design methods.

1 *The Nature of Design*

Design Activities

People have always designed things. One of the most basic characteristics of human beings is that they make a wide range of tools and other artefacts to suit their own purposes. As those purposes change, and as people reflect on the currently-available artefacts, so refinements are made to the artefacts, and completely new kinds of artefacts are conceived and made.

The wish to design things is therefore inherent in human beings, and 'designing' is not something that has always been regarded as needing special abilities. In traditional, craft-based societies 'designing' is not really separate from 'making'; that is to say, there is usually no prior activity of drawing or modelling before the activity of making the artefact. For example, a potter will make a pot by working directly with the clay, and without first making any sketches or drawings of the pot.

In modern, industrial societies, however, the activities of designing and of making artefacts are usually quite separate. The process of making something cannot normally start before the process of designing it is complete. In some cases—for example, in the electronics industry—the period of designing can take several years, whereas the average period of making each individual artefact might be measured only in hours or minutes.

Perhaps a way towards understanding this modern design activity is to begin at the end; to work backwards from the point where designing is finished and making can start. If making cannot start before designing is finished, then at least it is clear what the design process has to achieve. It has to provide a description of the artefact that is to be made. In this design description, almost nothing is left to the discretion of those involved in the process of making the artefact—it is specified down to the most detailed dimensions, to the kinds of surface finishes, to the materials, their colours, and so on.

In a sense, perhaps it does not matter how the designer works, so long as he or she produces that final description of the proposed artefact. When a client asks a designer for 'a design', that is what they want—the description. The focus of all design activities is that end-point.

Communication of Designs

The most essential design activity, therefore, is the production of a final description of the artefact. This has to be in a form that is understandable to those who will make the artefact. For this reason, the most widely-used form of communication is the drawing. For a simple artefact, such as a door-handle, one drawing would probably be enough, but for a larger, more complicated artefact such as a whole building the number of drawings may well run into hundreds, and for the most complex artefacts, such as chemical process plants, aeroplanes or major bridges, then thousands of drawings may be necessary.

These drawings will range from rather general descriptions— such as plans, elevations, and general arrangement drawings— that give an 'overview' of the artefact, to the most specific— such as sections and details—that give precise instructions on how the artefact is to be made. Because they have to communicate precise instructions, with minimal likelihood of misunderstanding, all the drawings are themselves subject to agreed rules, codes and conventions. These codes cover aspects such as how to lay out on one drawing the different views of an artefact relative to each other, how to indicate different kinds of materials, and how to specify dimensions. Learning to read and to make these drawings is an important part of design education.

The drawings will often contain annotations of additional information. Dimensions are one such kind of annotation. Written instructions may also be added to the drawings, such as notes on the materials to be used (as in Figure 1).

Other kinds of specifications as well as drawings may also be required. For example, the designer is often required to produce lists of all the separate components and parts that will make up the complete artefact, and an accurate count of the numbers of each component to be used. Written specifications of the standards of workmanship or quality of manufacture may also be necessary. Sometimes, an artefact is so complex, or so unusual,

Figure 1 Communication: a typical example of a conventional, engineering design detail drawing

that the designer makes a complete, three-dimensional mock-up or prototype version in order to communicate the design.

However, there is no doubt that drawings are the most useful form of communication of the description of an artefact that has yet to be made. Drawings are very good at conveying an understanding of what the final artefact has to be like, and that understanding is essential to the person who has to make the artefact.

Nowadays it is not always a person who makes the artefact; some artefacts are made by machines that have no direct human operator. These machines might be fairly sophisticated robots, or just simpler, numerically-controlled tools such as lathes or milling machines. In these cases, therefore, the final specification of a design prior to manufacture might not be in the form of drawings but in the form of a string of numbers represented on magnetic tape, or a computer program that controls the machine's actions.

It is therefore possible to imagine a design process in which no final communication drawings are made. The designer might be able to make a full-size or scale model of the artefact, and measurements could then be taken directly from the model and transmitted to numerically-controlled machines.

Evaluation of Designs

For the foreseeable future, however, drawings of various kinds will still be used elsewhere in the design process. Even if the final description is to be in the form of a string of numbers, the designer will probably want to make drawings for other purposes.

One of the most important of these other purposes is the checking, or evaluating, of design proposals before deciding on a final version for manufacture. The whole point of having the process of design separated from the process of making is that proposals for new artefacts can be checked before they are put into production. At its simplest, the checking procedure might merely be concerned with, say, ensuring that different components will fit together in the final design; this is an attempt to foresee possible errors and to ensure that the final design is workable. More complicated checking procedures might be concerned with, say, analyzing the forces in a proposed design to ensure that each component is designed to withstand the loads on it (Figure 2); this involves a process of refining a design to

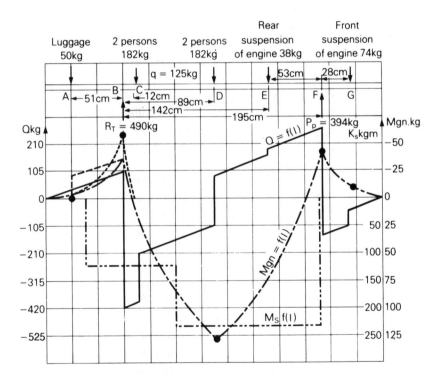

Figure 2
Evaluation: calculation of
the shear forces and
bending moments in the
body of the small
automobile

meet certain criteria such as maximum strength, or minimum weight or cost.

This process of refinement can be very complicated and can be the most time-consuming part of the design process. Imagine, for example, the design of a bridge. The designer must first propose the form of the bridge and the materials of which it will be made. In order to check that the bridge is going to be strong enough and stiff enough for the loads that it will carry, the designer must analyze the structure to determine the ways in which loads will be carried by it, what those loads will be in each member of the structure, what deflections will occur, and so on. After a first analysis, the designer might realize—or at least, suspect—that changing the locations or angles of some members in the bridge will provide a more efficient distribution of loadings throughout the whole structure. But these changes will mean that the whole structure will have to be re-analyzed and the loads recalculated.

In this kind of situation, it can be easy for the designer to become trapped in an iterative loop of decision-making, where improvements in one part of the design lead to adjustments in another part which lead to problems in yet another part. These problems may mean that the earlier 'improvement' is not feasible. This *iteration* is a common feature of designing.

Nevertheless, despite these potential frustrations, this process of refinement is a key part of designing. It consists, firstly, of analyzing a proposed design, and for this the designer needs to apply a range of engineering science or other knowledge. In many cases, specialists with more expert knowledge are called in to carry out these analyses. Then, secondly, the results of the analysis are evaluated against the design criteria—does the design come within the cost limit?, does it have enough space within it?, does it meet the minimum strength requirements?, does it use too much fuel?, and so on. In some cases, such criteria are set by government regulations, or by industry standards; others are set by the client or customer.

Many of the analyses are numerical calculations, and therefore again it is possible to imagine that drawings might not be necessary. However, specialists who are called in to analyze certain aspects of the design will almost certainly want a drawing—or other model of the design—before they can start work. Visualizations of the proposed design may also be important for the client and designer to evaluate aspects such as appearance, form and colour.

*Generation of
Designs*

Before any of these analyses and evaluations can be carried out the designer must, of course, first generate a design proposal. This is often regarded as the mysterious, creative part of designing—the client makes what might well be a very brief statement of requirements, and the designer responds (after a suitable period of time) with a design proposal, as if from nowhere. In reality, the process is less 'magical' than it appears.

In most cases, for instance, the designer is asked to design something similar to that which he or she has designed before, and therefore there is a stock of previous design ideas on which

Figure 3
Generation: early sketch design drawings for a small house by the architect Charles Moore

to draw. In some cases, only minor modifications are required to a previous design.

Nevertheless, there is something mysterious about the human ability to propose a design for a new—or even just a modified—artefact. It is perhaps as mysterious as the human ability to speak a new sentence—whether it is completely new, or just a modification of one heard, read, or spoken before.

This ability to design depends partly on being able to visualize something internally, in 'the mind's eye', but perhaps it depends even more on being able to make external visualizations. Once again, drawings are a key feature of the design process. At this early stage of the process, the drawings that the designer makes are not usually meant to be communications to anyone else. Essentially, they are communications with oneself—a kind of thinking aloud. As the example of early concept sketches for a house design shows (Figure 3), at this stage the designer is thinking of many aspects together, such as plan, elevation, section, structure, modular dimensions, materials, shape, overall form and function.

Exploration of Designs

At the start of the design process, the designer is usually faced with a very poorly defined problem; yet he or she has to come up with a well-defined solution. If one thinks of the problem as a territory, then it is largely unexplored and unmapped, and perhaps imaginary in places! As J. C. Jones has suggested, and as will be discussed in Chapter 11, it is therefore appropriate to think of the designer as an explorer, searching for the 'buried treasure' of a satisfactory solution concept.

Equally, if one thinks of all potential solutions as occupying a kind of solution space, then that, too, is relatively undefined—and perhaps infinite. The designer's difficulties are therefore two-fold: understanding the problem and finding a solution.

Often these two complementary aspects of design—problem and solution—have to be developed side-by-side. The designer makes a solution proposal and uses that to help understand what the problem 'really' is and what appropriate solutions might be like. The very first conceptualizations and representations of problem and solution are therefore critical to the kinds of searches and other procedures that will follow, and so to the final solution that will be designed.

The exploration of design solution-and-problem is also often done through early sketching of tentative ideas. It is necessary

Figure 4
Exploration: an example
of problem and solution
being explored together
for the 'Africar'—a simple
but robust automobile
suitable for conditions in
developing countries

because normally there is no way of directly generating an 'optimum' solution from the information provided in the design brief. Quite apart from the fact that the client's brief to the designer may be rather vague, there will be a wide range of criteria to be satisfied, and probably no single objective that must be satisfied above all others (Figure 4).

THE COMPOSITE SOLAR ELECTRIC DIESEL CAR

Design Problems

Design problems normally originate as some form of problem statement provided to the designer by someone else—the client or the company management. These problem statements—normally called a design 'brief'—can vary widely in their form and content. At one extreme, they might be something like the statement made by President Kennedy in 1961, setting a goal for the USA, 'before the end of the decade, to land a man on the moon and bring him back safely'. In this case, the goal was fixed, but the means of achieving it were very uncertain. The only constraint in the 'brief' was one of time—'before the end of the decade'. The designers were given a completely novel problem, a fixed goal, only one constraint, and huge resources of money, materials and people. This is quite an unusual situation for designers to find themselves in!

At the other extreme is the example of the brief provided to the industrial designer Eric Taylor, for an improved pair of photographic darkroom forceps. According to Taylor, the brief originated in a casual conversation with the company's Managing Director, who said to him, 'I was using these forceps last night, Eric. They kept slipping into the tray. I think we could do better than that.' In this case, the brief implied a design modification to an existing product, the goal was rather vague—'that [they] don't slip into the tray'—and the resources available to the designer would have been very limited for such a low-cost product.

Somewhere between these extremes would fall the more normal kind of design brief. A typical example might be the brief provided to the design department by the planning department of a company manufacturing plumbing fittings, shown on the next page. It is for a domestic hot and cold water mixing tap that can be operated with one hand.

What all of these three examples of design problems have in common is that they set a *goal*, some *constraints* within which the goal must be achieved, and some *criteria* by which a successful solution might be recognized. They do not specify what the solution will be, and there is no certain way of proceeding from the statement of the problem to a statement of the solution—except by 'designing'. Unlike some other kinds of problems, the person setting the problem does not know what 'the answer' is—but they will recognize it when they see it.

Even this last statement is not always true; sometimes clients do not 'recognize' the design solution when they see it. A famous example of early Modern Architecture is the 'Tugendhat

---

One-handed water mixing tap

Required: one-handed household water mixing tap with the following characteristics:

Throughput	10 l/min
Maximum pressure	6 bar
Normal pressure	2 bar
Hot water temperature	60°C
Connector size	10 mm

Attention to be paid to appearance. The firm's trade mark to be prominently displayed. Finished product to be marketed in two years' time. Manufacturing costs not to exceed DM 30 each at a production rate of 3000 taps per month.

House' in Germany, designed in the 1920s by Mies van der Rohe. Apparently the client had approached the architect after seeing some of the rather more conventional houses that he had designed. According to van der Rohe, when he showed the surprising, new design to the client, 'He wasn't very happy at first. But then we smoked some good cigars, . . . and we drank some glasses of a good Rhein wine, . . . and then he began to like it very much.'

So the solution that the designer generates may be something that the client 'never imagined might be possible', or perhaps even 'never realized was what they wanted'. Even a fairly precise problem statement gives no indication of what a solution *must* be. It is this uncertainty that makes designing such a challenging activity.

Ill-defined Problems　　The kinds of problems that designers tackle are regarded as 'ill-defined' or 'ill-structured', in contrast to well-defined or well-structured problems such as chess-playing, crossword puzzles or standard calculations. *Well-defined* problems have a clear goal, often one correct answer, and rules or known ways of proceeding that will generate an answer. The characteristics of *ill-defined* problems can be summarized as follows:

1. There is no definitive formulation of the problem
When the problem is initially set, the goals are usually vague, and many constraints and criteria are unknown. The problem context is often complex and messy, and poorly understood. In the course of problem-solving, temporary formulations of the problem may be fixed, but these are unstable and can change as more information becomes available.

2. Any problem formulation may embody inconsistencies
The problem is unlikely to be internally consistent; many conflicts and inconsistencies have to be resolved in the solution. Often, inconsistencies emerge only in the process of problem-solving.

3. Formulations of the problem are solution-dependent
Ways of formulating the problem are dependent upon ways of solving it; it is difficult to formulate a problem statement without implicitly or explicitly referring to a solution concept. The way the solution is conceived influences the way the problem is conceived.

4. Proposing solutions is a means of understanding the problem
Many assumptions about the problem, and specific areas of uncertainty, can be exposed only by proposing solution concepts. Many constraints and criteria emerge as a result of evaluating solution proposals.

5. There is no definitive solution to the problem
Different solutions can be equally valid responses to the initial problem. There is no objective true-or-false evaluation of a solution; but solutions are assessed as good or bad, appropriate or inappropriate.

Design problems are widely recognized as being ill-defined problems. It is usually possible to take some steps towards improving the initial definition of the problem, by questioning the client, collecting data, carrying out research, etc. There are also some rational procedures and techniques which can be applied in helping to solve ill-defined problems. But the designer's traditional approach, as suggested in some of the statements about ill-defined problems listed above, is to try to move fairly quickly to a potential solution, or set of potential solutions, and to use that as a means of further defining and understanding the problem.

Problem Structure

However, even when the designer has progressed well into the definition of a solution, difficulties in the problem structure may well still come to light. In particular, sub-solutions can be found to be interconnected with each other in ways that imply a 'pernicious' structure to the problem, e.g. a sub-solution that resolves a particular sub-problem may create irreconcilable conflicts with other sub-problems.

An example of this 'pernicious' problem structure was found in a study of housing design by Luckman. The architects identified five decision areas, or sub-problems, concerned with the directions of span of the roof and first floor joists, and the provision of load-bearing or non-load-bearing walls and partitions at ground- and first-floor levels. Making a decision in one area (say, direction of roof span) had implications for the first-floor partitions, and therefore the ground-floor partitions, which had implications for the direction of span of first-floor joists, and therefore for which of the external walls would have to be designed to be load-bearing. This had implications not only for the design of the external wall elevations, but also for the direction of span of the roof; and so they came full-circle back to the first decision area. This problem structure is shown diagrammatically in Figure 5, illustrating the cyclical structure that is often found in design problems.

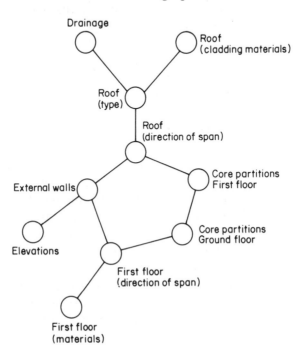

Figure 5
Problem structure in a housing design problem

As part of the research study, the individual sub-solution options in each decision area were separated out and the incompatible pairs of options identified. With this approach, it was possible to enumerate all the feasible solutions (i.e. sets of options containing no incompatible pairs). There were found to be eight feasible solutions, and relative costings of each could indicate which would be the cheapest solution.

This example shows that a rigorous approach can sometimes be applied even when the problem appears to be ill-defined, and the problem structure pernicious. This lends some support to those who argue that design problems are not always as ill-defined or ill-structured as they might appear to be. In particular, Simon has argued that there is no very clear boundary between 'ill-structured' and 'well-structured' problems, and that problems that have been regarded as 'ill-structured' can sometimes, by research and analysis, be reformulated as 'well-structured' problems. However, research into the behaviour of designers has shown that they will often treat a given problem *as though* it is ill-structured, even when it is presented as a well-structured problem, in order that they can create something innovative.

Problem-solving Strategies

Research has also shown that designers often attempt to avoid cycling around the pernicious decision loops of design problems by making high-level strategic decisions about solution options. Having identified a number of options, the designer selects what appears to be the best one for investigation at a more detailed level; again, several options are usually evident, and again the best is chosen. This results in what is known as a 'decision tree', with more and more branches opening from each decision point. An example is shown in Figure 6, based on a study by Marples of a valve design problem for a nuclear reactor. From ten different identified valve arrangements, one was selected for further development. This threw up a number of sub-problems, each of which had alternative solution possibilities; the best was selected, and so on.

This hierarchical 'top-down' approach to design is quite common, although sometimes a 'bottom-up' approach is used, starting with the lowest-level details and building up to a complete overall solution concept. The decision tree approach perhaps implies that the result is the best possible design, since the best options are chosen at each level. However, a decision at any

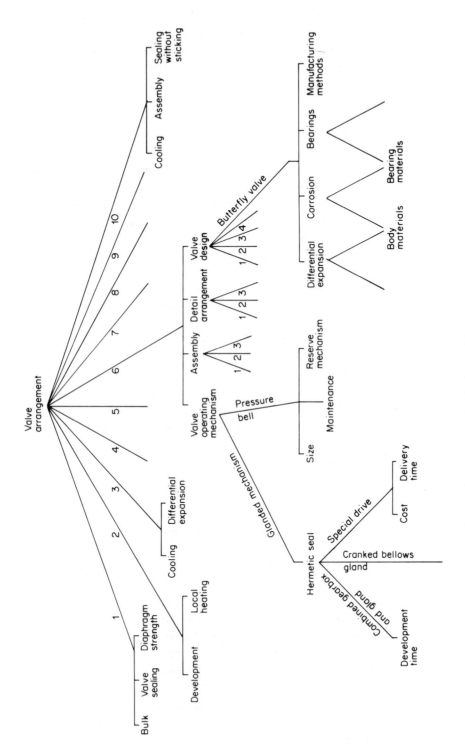

Figure 6 Part of the problem structure in a nuclear reactor valve design problem

particular level may well turn out to be sub-optimal in the light of subsequent options available at the other levels. For this reason, there is frequent back-tracking up and down the levels of hierarchy in the decision tree.

Design Ability

The world is full of tools, utensils, machines, buildings, furniture, clothes and so many other things that human beings apparently need or want in order to make their lives better. In fact, everything around us that isn't a simple piece of Nature has been designed by someone. Even a blank piece of paper has had design decisions made about its size, colour, density, opacity, absorbency, and so on.

Although there is so much design activity going on in the world, the ways in which people design are actually rather poorly understood. Until recently, design ability has been taken for granted as something that many people have, and that a few people have a particular talent in. However, there is now a growing body of knowledge about the nature of designing, about design ability and how to develop it, and about the design process and how to improve it.

When designers are asked to discuss their abilities, and to explain how they work, a few common themes emerge. One theme is the importance of creativity and 'intuition' in design—even in engineering design. For example, the engineering designer Jack Howe has said:

> I believe in intuition. I think that's the difference between a designer and an engineer. . . . I make a distinction between engineers and engineering designers. . . . An engineering designer is just as creative as any other sort of designer.

Some rather similar comments have been made by the industrial designer Richard Stevens:

> A lot of engineering design is intuitive, based on subjective thinking. But an engineer is unhappy doing this. An engineer wants to test; test and measure. He's been brought up this way and he's unhappy if he can't prove something. Whereas an industrial designer, with his Art School training, is entirely happy making judgements which are intuitive.

Another theme that emerges from designers' own comments is based on the recognition that problems and solutions in design are closely interwoven; that 'the solution' isn't always a straightforward answer to 'the problem'. For example, the furniture designer Geoffrey Harcourt commented on one of his creative designs like this:

> As a matter of fact, the solution that I came up with wasn't a solution to the problem at all. I never saw it as that. . . . But when the chair was actually put together [it] in a way quite well solved the problem, but from a completely different angle, a completely different point of view.

A third common theme to emerge is the need to use sketches, drawings or models of various kinds as a way to explore the problem and solution together, and simply to make some progress when faced with the complexity of design. For example, Jack Howe has said that, when uncertain how to proceed:

> I draw something. Even if it's 'potty', I draw it. The act of drawing seems to clarify my thoughts.

Given these insights into the nature of design, it is not surprising that the engineering designer Ted Happold should suggest that:

> I really have, perhaps, one real talent; that is that I don't mind at all living in the area of total uncertainty.

The quotations above are taken from interviews conducted with a number of successful and eminent designers by Davies. Their comments support some of the hypotheses that have emerged from more objective observational studies of designers at work, and other research that has been conducted into the nature of design. Some of this research supports the view that designers have a particular, 'designerly' way of thinking and working.

Design Thinking

A research study by Lawson compared the ways in which designers (in this case architects) and scientists solved the same problem. The scientists tended to use a strategy of systematically trying to understand the problem, in order to look for underlying rules which would enable them to generate an optimum solution. In contrast, the designers tended to make initial explorations and then suggest a variety of possible solutions until they found one that was good, or satisfactory. The evidence from the experiments suggested that scientists problem-solve by analysis, whereas designers problem-solve by synthesis; scientists use 'problem-focused' strategies and designers use 'solution-focused' strategies.

The problem-solving strategies used by designers probably reflect the nature of the problems that they normally tackle. These problems cannot be stated sufficiently explicitly such that solutions can be derived directly from them. The designer has to take the initiative in finding a starting point and suggesting tentative solution areas. 'Solution' and 'problem' are then both developed in parallel, sometimes leading to a creative redefinition of the problem, or to a solution that lies outside the boundaries of what was assumed to be possible.

The solution-focused strategies of designers are perhaps the best way of tackling design problems, which are by nature ill-defined. In order to cope with the uncertainty of ill-defined problems, the designer has to have the self-confidence to define, redefine and change the problem as given, in the light of solutions that emerge in the very process of designing. People who seek the certainty of structured, well-defined problems will never appreciate the delight of being a designer!

2 *The Design Process*

Descriptive Models

There have been many attempts to draw up maps or models of the design process. Some of these models simply *describe* the sequences of activities that typically occur in designing; other models attempt to *prescribe* a better or more appropriate pattern of activities.

Descriptive models of the design process usually emphasize the importance of generating a solution concept early in the process, thus reflecting the 'solution-focused' nature of design thinking. This initial solution 'conjecture' is then subjected to analysis, evaluation, refinement and development. Sometimes, of course, the analysis and evaluation show up fundamental flaws in the initial conjecture and it has to be abandoned, a new concept generated and the cycle started again. The process is *heuristic*: using previous experience, general guidelines and 'rules of thumb' that lead in what the designer hopes to be the right direction, but with no absolute guarantee of success.

In Chapter 1, I developed a simple descriptive model of the design process, based on the essential activities that the designer performs. The end-point of the process is the *communication* of a design, ready for manufacture. Prior to this, the design proposal is subject to *evaluation* against the goals, constraints and criteria of the design brief. The proposal itself arises from the *generation* of a concept by the designer, usually after some initial *exploration* of the ill-defined problem space. Putting these four activity types in their natural sequence, we have a simple four-stage model of the design process consisting of exploration—generation—evaluation—communication.

This simple four-stage model is shown diagrammatically in Figure 7. Assuming that the evaluation stage does not always lead directly onto the communication of a final design, but that

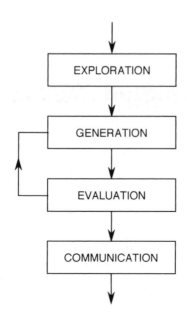

Figure 7
A simple four-stage
model of the design
process

sometimes a new, more satisfactory concept has to be chosen,
an iterative feedback loop is shown from the evaluation stage to
the generation stage.

Models of the design process are often drawn in this flow-
diagram form, with the development of the design proceeding
from one stage to the next, but with feedback loops showing
the iterative returns to earlier stages which are frequently necessary. For example, French has developed a more detailed model
of the design process, shown in Figure 8, based on the following
activities:

Analysis of problem
Conceptual design
Embodiment of schemes
Detailing

In the diagram, the circles represent stages reached, or outputs,
and the rectangles represent activities, or work in progress.

The process begins with an initial statement of a 'need', and
the first design activity is 'analysis of the problem'. French
suggests that:

> The analysis of the problem is a small but important part of the
> overall process. The output is a statement of the problem, and this
> can have three elements,

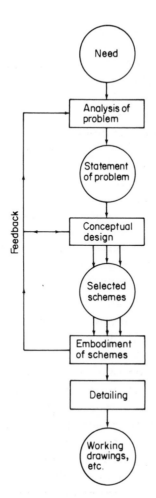

Figure 8
French's model of the
design process

1. A statement of the design problem proper
2. Limitations placed upon the solution, e.g. codes of practice, statutory requirements, customers' standards, date of completion, etc.
3. The criterion of excellence to be worked to.

These three elements correspond to the goals, constraints and criteria of the design brief.

The activities that follow, according to French, are then:

Conceptual design
This phase . . . takes the statement of the problem and generates broad solutions to it in the form of schemes. It is the phase that makes the greatest demands on the designer, and where there is the most scope for striking improvements. It is the phase where

engineering science, practical knowledge, production methods, and commercial aspects need to be brought together, and where the most important decisions are taken.

Embodiment of schemes

In this phase the schemes are worked up in greater detail and, if there is more than one, a final choice between them is made. The end product is usually a set of general arrangement drawings. There is (or should be) a great deal of feedback from this phase to the conceptual design phase.

Detailing

This is the last phase, in which a very large number of small but essential points remains to be decided. The quality of this work must be good, otherwise delay and expense or even failure will be incurred; computers are already reducing the drudgery of this skilled and patient work and reducing the chance of errors, and will do so increasingly.

These activities are typical of conventional engineering design. Figures 9, 10 and 11 illustrate the type of work that goes on in each stage. The illustrations are examples from the design of the small concrete mixer. *Conceptual design* is shown in Figure 9, where three alternatives are proposed for the drive connection from the motor to the mixing drum. *Embodiment design* is shown in Figure 10, where concept (c) is developed in terms of how to support and assemble the motor, drum, pulleys, etc. Figure 11 shows a small example of *detail design*, in which the motor mounting plate is redesigned from a welded T-shape to a channel section U-shape, after tests of a prototype found excessive vibration occurring in the original.

Figure 9
Conceptual design:
alternatives for the drive
connection in a small
concrete mixer

Figure 10 Embodiment design: one concept developed in more detail

Figure 11 Detail design: redesign of a mounting plate to resist vibration

Prescriptive Models

As well as models that simply describe a more-or-less conventional, heuristic process of design, there have been several attempts at building prescriptive models of the design process. These latter models are concerned with trying to persuade or encourage designers to adopt improved ways of working. They usually offer a more *algorithmic*, systematic procedure to follow, and are often regarded as providing a particular *design methodology*.

Many of these prescriptive models have emphasized the need for more analytical work to precede the generation of solution concepts. The intention is to try to ensure that the design problem is fully understood, that no important elements of it are overlooked, and that the 'real' problem is identified. There are plenty of examples of excellent solutions to the wrong problem!

These models have therefore tended to suggest a basic structure to the design process of analysis—synthesis—evaluation. These stages were defined by Jones in an early example of a systematic design methodology as follows:

> *Analysis.* Listing of all design requirements and the reduction of these to a complete set of logically related performance specifications.
>
> *Synthesis.* Finding possible solutions for each individual perform-ance spec-ification and building up complete designs from these with least possible compromise.
>
> *Evaluation.* Evaluating the accuracy with which alternative designs fulfil performance requirements for operation, manufacture and sales before the final design is selected.

This may sound very similar to a conventional design process, but the emphases here are on performance specifications logically derived from the design problem, generating several alternative design concepts by building-up the best sub-solutions and making a rational choice of the best of the alternative designs. Such apparently sensible, rational procedures are not always followed in conventional design practice.

A more detailed prescriptive model was developed by Archer, and is summarized in Figure 12. This includes interactions with the world outside of the design process itself, such as inputs from the client, the designer's training and experience, other sources of information, etc. The output is, of course, the communication of a specific solution. These various inputs and

Figure 12
Archer's model of the
design process

outputs are shown as external to the design process in the flow diagram, which also features many feedback loops.

Within the design process, Archer identified six types of activity:

> *Programming*—establish crucial issues; propose a course of action.
>
> *Data collection*—collect, classify and store data.
>
> *Analysis*—identify sub-problems; prepare performance (or design) specifications; reappraise proposed programme and estimate.
>
> *Synthesis*—prepare outline design proposals.
>
> *Development* —develop prototype design(s); prepare and execute validation studies.
>
> *Communication*—prepare manufacturing documentation.

Archer summarized this process as dividing into three broad phases: analytical, creative and executive (Figure 13). He sug-

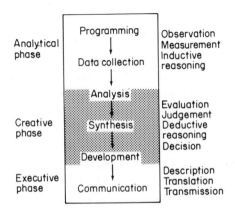

Figure 13
Archer's three-phase
summary model of the
design process

gested that:

> One of the special features of the process of designing is that the
> analytical phase with which it begins requires objective observa-
> tion and inductive reasoning, while the creative phase at the heart
> of it requires involvement, subjective judgement, and deductive
> reasoning. Once the crucial decisions are made, the design pro-
> cess continues with the execution of working drawings, schedu-
> les, etc., again in an objective and descriptive mood. The design
> process is thus a creative sandwich. The bread of objective and
> systematic analysis may be thick or thin, but the creative act is
> always there in the middle.

Some much more complex models have been proposed, but
they often tend to obscure the general structure of the design
process by swamping it in the fine detail of the numerous tasks
and activities that are necessary in all practical design work. A
reasonably comprehensive model that still retains some clarity
is that offered by Pahl and Beitz (Figure 14). It is based on the
following design stages:

> *Clarification of the task*
> Collect information about the requirements to be embodied in the
> solution and also about the constraints.
>
> *Conceptual design*
> Establish function structures; search for suitable solution princi-
> ples; combine into concept variants.
>
> *Embodiment design*
> Starting from the concept, the designer determines the layout and
> forms and develops a technical product or system in accordance
> with technical and economic considerations.
>
> *Detail design*
> Arrangement, form, dimensions and surface properties of all the
> individual parts finally laid down; materials specified; technical
> and economic feasibility re-checked; all drawings and other pro-
> duction documents produced.

Considerable work on these kinds of models and on other
aspects of rationalizing the design process has been done in
Germany. The professional engineers' body, Verein Deutscher
Ingenieure (VDI), has produced a number of 'VDI-Guidelines'
in this area, including VDI 2221, 'Systematic Approach to the
Design of Technical Systems and Products'. This Guideline
suggests a systematic approach in which 'The design process,
as part of product creation, is subdivided into general working

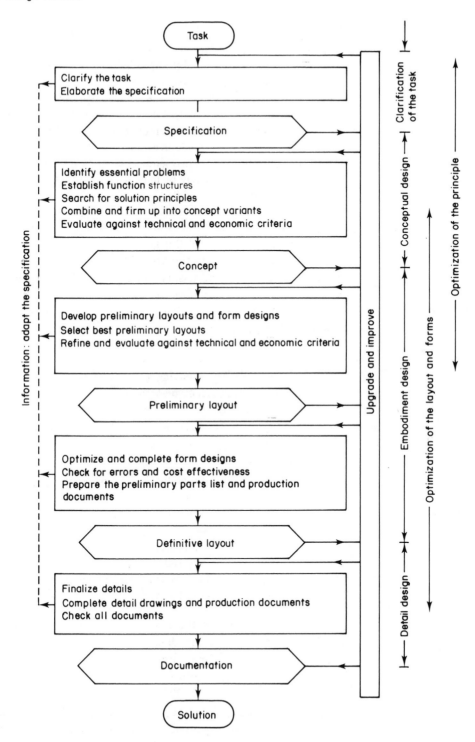

Figure 14   Pahl and Beitz's model of the design process

stages, making the design approach transparent, rational and independent of a specific branch of industry'.

The structure of this general approach to design is shown in Figure 15, and is based on seven stages, each with a particular output. The output from the first stage, the specification, is regarded as particularly important, and is constantly reviewed and kept up-to-date, and used as a reference in all the subsequent stages.

Figure 15
The VDI 2221 model of the design process

The second stage of the process consists of determining the required functions of the design, and producing a diagrammatic

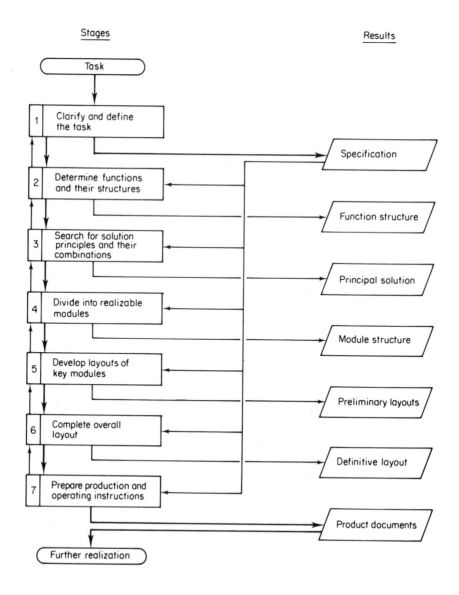

function structure. In stage 3 a search is made for solution principles for all sub-functions, and these are combined in accordance with the overall function structure into a principal solution. This is divided, in stage 4, into realizable modules and a module structure representing the breakdown of the solution into fundamental assemblies. Key modules are developed in stage 5 into a set of preliminary layouts. These are refined and developed in stage 6 into a definitive layout, and the final product documents are produced in stage 7.

In the Guideline it is emphasized that several solution variants should be analyzed and evaluated at each stage, and that there is a lot more detail in each stage than is shown in the diagram. The following words of warning about the approach are also given:

> It is important to note that the stages do not necessarily follow rigidly one after the other. They are often carried out iteratively, returning to preceding ones, thus achieving a step-by-step optimization.

The VDI-Guideline follows a general systematic procedure of first analyzing and understanding the problem as fully as possible, then breaking this into sub-problems, finding suitable sub-solutions and combining these into an overall solution. The procedure is shown diagrammatically in Figure 16.

This kind of procedure has been criticized in the design world because it seems to be based on a problem-focused, rather than a solution-focused approach. It therefore runs counter to the designer's traditional ways of thinking.

A more radical model of the design process, which recognizes the solution-focused nature of design thinking, has been suggested by March (Figure 17). He argued that the two conventionally-understood forms of reasoning—inductive and deductive—only apply logically to the evaluative and analytical types of activity in design. But the type of activity that is most particularly associated with design is that of synthesis, for which there is no commonly-acknowledged form of reasoning. March drew on the work of the philosopher C. S. Peirce to identify this missing concept of 'abductive' reasoning. According to Peirce,

> Deduction proves that something *must be*; induction shows that something *actually is* operative; abduction merely suggests that something *may be*.

Figure 16
The VDI 2221 model of
development from
problem to solution

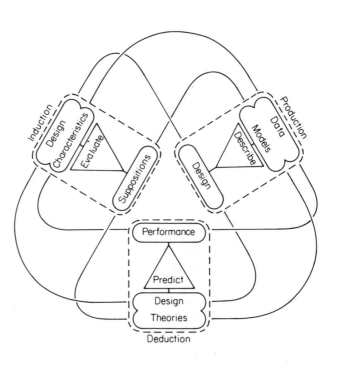

Figure 17
March's model of the
design process

It is this hypothesizing of what *may* be, the act of synthesis, that is central to design. Because it is the kind of thinking by which designs are generated or produced, March prefers to call it 'productive' reasoning. Thus his model for a rational design process is a 'PDI model': Production – Deduction – Induction.

In this model, the first phase, of *productive* reasoning, draws on a preliminary statement of requirements, and some presuppositions about solution types in order to produce, or describe, a design proposal. From this proposal and established theory (e.g. engineering science) it is possible *deductively* to analyze, or predict, the performance of the design. From these predicted performance characteristics it is possible *inductively* to evaluate further suppositions or possibilities, leading to changes or refinements in the design proposal.

Certainly it seems that, in most design situations, it is not possible, or relevant, to attempt to analyze 'the problem' *ab initio* and in abstract isolation from solution concepts; the designer needs to explore and develop problem and solution together. Although there may be some logical progression from problem to sub-problems and from sub-solutions to solution, there is a symmetrical, commutative relationship between problem and solution, and between sub-problems and sub-solutions, as illustrated in Figure 18. The general direction of movement in this model is anticlockwise, but there are substantial periods of iterative activity, going to-and-fro between problem and solution.

Figure 18
The symmetrical relationships of problem/sub-problems/sub-solutions/solution in design

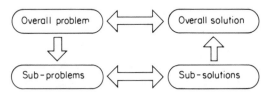

Systematic Procedures

There may be differences in their preferred models, but the proponents of systematic procedures all agree that there is an urgent need to improve on traditional ways of working in design.

There are several reasons for this concern to develop new design procedures. One is the increasing complexity of modern design. A great variety of new demands is increasingly being made on the designer, such as the new materials and devices (e.g. electronics) that become available and the new problems that are presented to designers. Many of the products and machines to be designed today have never existed before, and so the designer's previous experience may well be irrelevant and inadequate for these tasks. Therefore a new, systematic approach is needed, it is argued.

A related part of the complexity of modern design is the need to develop team work, with many specialists collaborating in and contributing to the design. To help coordinate the team, it is necessary to have a clear, systematic approach to design, so that specialists' contributions are made at the right point in the process. Dividing the overall problem into sub-problems in a systematic procedure also means that the design work itself can be sub-divided and allocated to appropriate team members.

As well as being more complex, modern design work often has very high risks and costs associated with it. For example, many products are designed for mass manufacture, and the costs of setting up the manufacturing plant, buying-in raw materials, and so on, are so high that the designer cannot afford to make mistakes: the design must be absolutely right before it goes into production. This means that any new product must have been through a careful process of design. Other kinds of large, one-off designs, such as chemical process plants, or complex products such as aeroplanes, also have to have a very rigorous design process to try to ensure their safe operation and avoid the catastrophic consequences of failure.

Finally, there is a more general concern with trying to improve the efficiency of the design process. In some industries there is a pressing need to ensure that the lead-time necessary to design a new product is kept to a minimum. In all cases, it is desirable to try to avoid the mistakes and delays that often occur in conventional design procedures. The introduction of computers already offers one way of improving the efficiency of the design process, and is also in itself an influence towards more systematic ways of working.

3 *Design Methods*

In a sense, any identifiable way of working, within the context of designing, can be considered to be a design method. The most common design method can be called the method of 'design-by-drawing'. That is to say, most designers rely extensively on drawing as their main aid to designing.

Design methods are any procedures, techniques, aids, or 'tools' for designing. They represent a number of distinct kinds of activities that the designer might use and combine into an overall design process. Although some design methods can be the conventional, normal procedures of design, such as drawing, there has been a substantial growth in recent years of new, unconventional procedures that are more usually grouped together under the name of 'Design Methods'.

The main intention of these new methods is that they attempt to bring rational procedures into the design process. Some of these methods are themselves new inventions, some are adapted from operational research, decision theory, management sciences or other sources, and some are simply extensions or formalizations of the informal techniques that designers have always used. For example, the informal methods of looking up manufacturers' catalogues or seeking advice from colleagues might be formalized into an 'information search' method; or informal procedures for saving costs by detailed redesigning of a component can be formalized into a 'value analysis' method.

It seems that some of these new methods can become *over*-formalized, or can be merely fancy names for old, commonsense techniques. They can also appear to be *too* systematic to be useful in the rather messy and often hurried world of the design office. For these kinds of reasons, many designers are still mistrustful of the whole idea of 'design methods'.

The counter-arguments to that view are based on the reasons for adopting systematic procedures which were outlined in Chapter 2. For instance, many modern design projects are too complex to be resolved satisfactorily by the old, conventional methods. There are also too many errors made with conventional ways of working, and they are not very useful where team work is necessary. Design methods try to overcome these kinds of problems, and—above all—they try to ensure that a better product results from the new design process.

What Kinds of Design Methods Are There?

There have been many new methods developed to help overcome the difficulties of modern design problems. For example, the textbook of *Design Methods* by Jones contains descriptions of 35 methods, including the following:

Methods of exploring design situations

Method	Aim
Stating objectives	To identify external conditions with which the design must be compatible
Literature searching	To find published information that can favourably influence the designers' output and that can be obtained without unacceptable cost and delay
Searching for visual inconsistencies	To find directions in which to search for design improvements
Interviewing users	To elicit information that is known only to users of the product or system in question
Questionnaires	To collect usable information from the members of a large population
Investigating user behaviour	To explore the behaviour patterns, and to predict the performance limits, of potential users of a new design
Data logging and data reduction	To infer, and to make visible, patterns of behaviour upon which critical design decisions depend

Methods of searching for ideas	**Method**	**Aim**
	Brainstorming	To stimulate a group of people to produce many ideas quickly
	Synectics	To direct the spontaneous activity of the brain and the nervous system towards the exploration and transformation of design problems
	Removing mental blocks	To find new directions of search when the apparent search space has yielded no wholly acceptable solution
	Morphological charts	To widen the area of search for solutions to a design problem

Methods of exploring problem structure	**Method**	**Aim**
	Interaction matrix	To permit a systematic search for connections between elements within a problem
	Interaction net	To display the pattern of connections between elements within a design problem
	Analysis of interconnected decision areas (AIDA)	To identify and to evaluate all the compatible sets of sub-solutions to a design problem
	System transformation	To find ways of transforming an unsatisfactory system so as to remove its inherent faults
	Functional innovation	To find a radically new design capable of creating new patterns of behaviour and demand
	Alexander's method of determining components	To find the right physical components of a physical structure such that each component can be altered independently to suit future changes in the environment
	Classification of design information	To split a design problem into manageable parts

Methods of evaluation

Method	**Aim**
Checklists	To enable designers to use knowledge of requirements that have been found to be relevant in similar situations
Selecting criteria	To decide how an acceptable design is to be recognized
Ranking and weighting	To compare a set of alternative designs using a common scale of measurement
Specification writing	To describe an acceptable outcome for designing that has yet to be done

As can be seen from the list, some methods are formal versions of conventional procedures (e.g. literature searching, interviewing users), some are applications of methods first developed in other fields (e.g. data logging from science; brainstorming from advertising), and some are new inventions (e.g. Alexander's method, AIDA). The list also shows that different design methods have different purposes and are relevant to different aspects of and stages in the design process.

Versions of several of the methods which are described in detail in this book, in Chapters 4–10, are included in the list above—stating objectives (objectives tree), morphological charts, ranking and weighting (weighted objectives) and specification writing (performance specification). Other methods which will also be described in detail are function analysis, quality function deployment and value engineering. Some of the creativity methods—brainstorming, synectics and removing mental blocks—will be described in outline later in this chapter.

What do Different Design Methods Have in Common?

Looking over all the methods in the above list, two principal common features emerge. One is that design methods *formalize* certain procedures of design, and the other is that design methods *externalize* design thinking. Formalization is a common feature of design methods because they attempt to avoid the occurrence of oversights, of overlooked factors in the design problem, of the kinds of errors that occur with informal methods. The process of formalizing a procedure also tends to widen the approach that is taken to a design problem and to widen the search for appropriate solutions—it encourages and enables you to think beyond the first solution that comes into your head.

This is also related to the other general aspect of design methods, that they externalize design thinking, i.e. they try to get your thoughts and thinking processes out of your head and into the charts and diagrams that commonly feature in design methods. This externalizing is a significant aid when dealing with complex problems, but it is also a necessary part of team work, i.e. providing means by which all the members of the team can see what is going on and can contribute to the design process. Getting a lot of systematic work out of your head and onto paper also means that your mind can be more free to pursue the kind of thinking it is best at—intuitive and imaginative thinking.

Design methods therefore are *not* the enemy of creativity, imagination and intuition. Quite the contrary: they are perhaps more likely to lead to novel design solutions than the informal, internal, and often incoherent thinking procedures of the conventional design process. Some design methods are, indeed, techniques specifically for aiding creative thought. In fact, the general body of design methods can be classified into two broad groups: creative methods and rational methods.

Creative Methods

There are several design methods which are intended to help stimulate creative thinking. In general, they work by trying to increase the flow of ideas, by removing the mental blocks that inhibit creativity, or by widening the area in which a search for solutions is made.

Brainstorming

The most widely-known creative method is brainstorming. This is a method for generating a large number of ideas, most of which will subsequently be discarded, but with perhaps a few novel ideas being identified as worth following up. It is normally conducted as a small group session of about four to eight people.

The group of people selected for a brainstorming session should be diverse. It should not just be experts or those knowledgeable in the problem area, but should include a wide range of expertise and even laypeople if they have some familiarity with the problem area. The group must be non-hierarchical, although one person does need to take an organizational lead.

The role of the group leader in a brainstorming session is to ensure that the format of the method is followed, and that it does not just degenerate into a round-table discussion. An important prior task for the leader is to formulate the problem statement used as a starting point. If the problem is stated too narrowly, then the range of ideas from the session may be rather limited. On the other hand, a very vague problem statement leads to equally vague ideas, which may be of no practical use. The problem can often be usefully formulated as a question, such as 'How can we improve on X?'.

In response to the initial problem statement, the group members are asked to spend a few minutes—in silence—writing down the first ideas that come into their heads. It is a good idea if each member has a pile of small record cards on which to write these and subsequent ideas. The ideas should be expressed succinctly, and written one per card.

The next, and major, part of the session is for each member of the group, in turn, to read out one idea from his or her set. The most important rule here is that *no criticism is allowed* from any other member of the group. The usual responses to unconventional ideas, such as 'That's silly', or 'That will never work', kill off spontaneity and creativity. At this stage, the feasibility or otherwise of any idea is not important—evaluation and selection will come later.

What each group member should do in response to every other person's idea is to try to build on it, to take it a stage further, to use it as a stimulus for other ideas, or to combine it with his or her own ideas. For this reason, there should be a short pause after each idea is read out, to allow a moment for reflection and for writing down further new ideas. However, the session must not become too stilted; the atmosphere should be relaxed and free-wheeling. A brainstorming session should also be fun: humour is often an essential ingredient of creativity.

The group session should not last more than about 20–30 minutes, or should be wound up when no more new ideas are forthcoming. The group leader, or someone else, then collects all the cards and spends a separate period evaluating the ideas. A useful aid to this evaluation is to sort or classify the ideas into related groups; this in itself often suggests further ideas, or indicates the major types of idea that there appear to be. If principal solution areas and one or two novel ideas result from a brainstorming session then it will have been worthwhile.

Participating in a brainstorming session is rather like playing a party game; and like a party game it only works well when everyone sticks to the rules. In fact, all design methods only work best when they are followed with some rigour, and not in a sloppy or half-hearted fashion. The essential rules of brainstorming are:

> No criticism is allowed during the session.
> A large quantity of ideas is wanted.
> Seemingly-crazy ideas are quite welcome.
> Keep all ideas short and snappy.
> Try to combine and improve on the ideas of others.

Example:
Container lock

This example shows how brainstorming can be applied to the task of creating a new solution to an old problem—the locking of containers (the large goods containers transported by lorries). The conventional solution is a padlock, but then the key for the padlock also has to be either transported together with the container (hence presenting an obvious security problem) or sent separately to the recipient (possibly getting lost). In practice, it seems that most container padlocks are 'opened' with a bolt-cutter, because no-one can find the key!

A short brainstorming session was held to generate ideas for solving this problem. The problem was stated as: 'Provide a means of securing containers that is tamper-proof but easy to open.' Within a few minutes, the following ideas were generated:

> Incorporate an electronic code—fax the code to the recipient
> Combination lock
> Time lock
> Clasps welded together
> A locked bolt that is easily cut to open it
> Padlocks with master keys retained by regular customers
> Giant stapler and staple-remover
> Ceramic bolt that can be smashed
> Glass bolt that sounds alarm when smashed
> Lorry driver swallows the key
> A 'puzzle' lock that can only be opened by a very skilled person

Some of these are fairly 'obvious' ideas, but getting them out of your head can sometimes seem to free the mental space for other ideas to come. Others are 'crazy' ideas, such as the lorry driver swallowing the key; in such a case, everyone knows where the key is, but has to wait a couple of days before it can be recovered! (Another sort of 'time lock' as the proposer explained!) There is also an example in the list of one idea building upon another: the 'Glass bolt that sounds an alarm when smashed' was a response to the 'Ceramic bolt' idea, but based also on fire alarm buttons that are activated by smashing the glass cover.

In reviewing this list of ideas several novel concepts come to mind, but perhaps most appealing is the simplicity of adapting what is already the unofficial but conventional solution—to cut the bolt off. A bolt could be made such that it was designed to be cut off. Made in two sections, the parts of the bolt would be pushed together to secure the container, and could only be opened by being cut. Colour coding and numbering each bolt would mean that it could not be replaced in transit, and if it was cut open then this would be obvious. Such simple but secure bolts would be cheaper than conventional padlocks.

The 'Oneseal' disposable lock is designed on these principles.

Synectics

Creative thinking often draws on analogical thinking—on the ability to see parallels or connections between apparently dissimilar topics. The role of humour is again relevant, since most jokes depend for their effect on the unexpected transfer or juxtaposition of concepts from one context to another, or what Koestler called the 'bisociation' of ideas. Bisociation plays a fundamental role in creativity.

The use of analogical thinking has been formalized in a creative design method known as 'synectics'. Like brainstorming, synectics is a group activity in which criticism is ruled out, and the group members attempt to build, combine and develop ideas towards a creative solution to the set problem. Synectics is different from brainstorming in that the group tries to work collectively towards a particular solution, rather than generating a large number of ideas. A synectics session is much longer than brainstorming, and much more demanding. In a synectics session, the group is encouraged to use particular types of analogy, as follows:

> *Direct analogies.* These are usually found by seeking a biological solution to a similar problem. For example, Brunel's observation of a shipworm forming a tube for itself as it bored through timber is said to have led him to the idea of a caisson for underwater constructions; 'Velcro' fastening was designed on an analogy with plant burrs.

Personal analogies. The team members imagine what it would be like to use oneself as the system or component that is being designed. For example, what would it feel like to be a motorcar suspension unit; how would I operate if I were a computerized filing system?

Symbolic analogies. Here, poetic metaphors and similes are used to relate aspects of one thing with aspects of another. For example, the 'friendliness' of a computer, the 'head' and 'claw' of a hammer, a 'tree' of objectives, the 'Greek key pattern' of a housing layout.

Fantasy analogies. These are 'impossible' wishes for things to be achieved in some 'magical' way. For example, 'What we really want is a doorkeeper who recognizes each system user'; 'We need the bumps in the road to disappear beneath the wheels.'

A synectics session starts with 'the problem as given'—the problem statement as presented by the client or company management. Analogies are then sought that help to 'make the strange familiar', i.e. expressing the problem in terms of some more familiar (but perhaps rather distant) analogy. This leads to a conceptualization of 'the problem as understood'—the key factor or elements of the problem that need to be resolved, or perhaps a complete reformulation of the problem. The problem as understood is then used to guide the use of analogies again, but this time to 'make the familiar strange'. Unusual, creative analogies are sought, which may lead to novel solution concepts. The analogies are used to open up lines of development which are pursued as hard and as imaginatively as possible by the group.

Example: Forklift truck

A design team looking for new versions of a company's forklift trucks focused on the problem area of using such trucks in warehouses for the stacking and removal of palleted goods. Conventional forklift trucks have to face head-on to the stacks in order to place and lift the pallets, and then be manoeuvred again within the aisle between the stacks in order to move to another location or to exit the warehouse. This means that the aisles have to be quite wide, using up warehouse space.

This example shows how synectics thinking can be used in the approach to such a problem. Direct analogies could be used to 'make the strange familiar', i.e. to familiarize the team with the new problem. For instance, analogies of the movement of snakes might be explored, leading to 'the problem as understood' being the need for a truck to twist sinuously in its manoeuvring. To 'make the familiar strange', the team might use personal and fantasy analogies of the kind: 'If I was holding the pallet in my outstretched arms, going along the aisle, I would like to be able to twist my upper body through 90 degrees (without moving my feet) to place the pallet in the rack.' Symbolic analogies of rotating turrets and articulated skeletons led eventually to a new design concept of an articulated truck with forks mounted on a front section that could swivel through 90 degrees.

The Translift 'Benditruck' is designed on these principles.

Enlarging The Search Space

A common form of mental block to creative thinking is to assume rather narrow boundaries within which a solution is sought. Many creativity techniques are aids to enlarging the 'search space'.

Transformation   One such technique attempts to 'transform' the search for a solution from one area to another. This often involves applying verbs that will transform the problem in some way, such as

> magnify, minify, modify, unify, subdue, subtract, add, divide, multiply, repeat, replace, relax, dissolve, thicken, soften, harden, roughen, flatten, rotate, rearrange, reverse, combine, separate, substitute, eliminate.

Random input   Creativity can be triggered by random inputs from whatever source. This can be applied as a deliberate technique, e.g. opening a dictionary or other book and choosing a word at random and using that to stimulate thought on the problem in hand. Or switch on a television set and use the first visual image as the random input stimulus.

Why? Why? Why?   Another way of extending the search space is to ask a string of questions 'why?' about the problem, such as 'Why is this device necessary?', 'Why can't it be eliminated?', etc. Each answer is followed up, like a persistent child, with another 'Why?' until a dead end is reached or an unexpected answer prompts an idea for a solution. There may be several answers to any particular 'Why?', and these can be charted as a network of question-and-answer chains.

Counter-planning This method is based on the concept of the dialectic, i.e. pitting an idea (the thesis) against its opposite (the antithesis) in order to generate a new idea (the synthesis). It can be used to challenge a conventional solution to a problem by proposing its deliberate opposite, and seeking a compromise. Alternatively, two completely different solutions can be deliberately generated, with the intention of combining the best features of each into a new synthesis.

The Creative Process

The methods above are some techniques which have been found useful when it is necessary for a designer or design team to 'turn on' their creative thinking. But creative, original ideas can also seem to occur quite spontaneously, without the use of any such aids to creative thinking. Is there, therefore, a more general process of creative thinking which can be developed?

Psychologists have studied accounts of creative thinking from a wide range of scientists, artists and designers. In fact, as most people have also experienced, these highly creative individuals generally report that they experience a very sudden creative insight that suggests a solution to the problem they have been working on. There is a sudden 'illumination'—just like the light-bulb flashing on that cartoonists use to suggest someone having a bright idea.

This creative 'Ah-ha!' experience often occurs when the individual is not expecting it, and after a period when they have been thinking about something else. This is rather like the common phenomenon of suddenly remembering a name or word that could not be recalled when it was wanted.

However, the sudden illumination of a bright idea does not usually occur without considerable background work on a problem. The illumination or key insight is also usually just the germ of an idea that needs a lot of further work to develop it into a proper, complete solution to the problem.

Similar kinds of thought sequence occur often enough in creative thinking for the psychologists to suggest that there is a general pattern to it. This general pattern is the sequence: Recognition—Preparation—Incubation—Illumination— Verification.

> *Recognition* is the first realization or acknowledgement that 'a problem' exists.

> *Preparation* is the application of deliberate effort to understand the problem.

Incubation is a period of leaving it to 'mull over' in the mind, allowing one's subconscious to go to work.

Illumination is the (often quite sudden) perception or formulation of the key idea.

Verification is the hard work of developing and testing the idea.

This process is essentially one of work–relaxation–work, with (if you are lucky) the creative insight occurring in a relaxation period. The hard work of preparation and verification is essential. Like most other kinds of creative activity, creative design is 1% inspiration and 99% perspiration!

Rational Methods

More commonly regarded as 'design methods' than the creativity techniques are the rational methods which encourage a systematic approach to design. Nevertheless, these rational methods often have similar aims to the creative methods, such as widening the search space for potential solutions, or facilitating team work and group decision-making. So it is not necessarily true that rational methods are somehow the very opposite of creative methods.

Many designers are suspicious of rational methods, fearing that they are a 'straitjacket', or that they are stifling to creativity. This is a misunderstanding of the intentions of systematic design, which is meant to improve the quality of design decisions, and hence of the end product. Creative methods and rational methods are complementary aspects of a systematic approach to design. Rather than a 'straitjacket', they should be seen as a 'lifejacket', helping the designer to keep afloat.

Perhaps the simplest kind of rational method is the checklist. Everyone uses this method in daily life—for example, in the form of a shopping list, or list of things to remember to do. It *externalizes* what you have to do, so that you don't have to try to keep it all in your head, and so that you don't overlook something. It *formalizes* the process, by making a record of items which can be checked-off as they are collected or achieved until everything is complete. It also allows team-work or participation by a wider group, e.g. all the family can contribute suggestions for the shopping list. It also allows sub-division of the task (i.e. improving the efficiency of the process), such as allocating separate sections of the list to different members of the team. In these respects, it is a model for most of the rational design methods. In design terms, a checklist may be a list of questions to be asked in the initial stages of design, or a list of features to be incorporated in the design, or a list of criteria, standards, etc. that the final design must meet.

There is a wide range of rational design methods, as we saw in the long list above, covering all aspects of the design process from problem clarification to detail design. The next seven chapters present a selection of the most relevant and widely-used methods, also covering the whole design process. The selected set is as follows:

Stage in the design process	Method relevant to this stage
Clarifying objectives	Objectives Tree Aim: To clarify design objectives and sub-objectives, and the relationships between them.
Establishing functions	Function Analysis Aim: To establish the functions required, and the system boundary, of a new design.
Setting requirements	Performance Specification Aim: To make an accurate specification of the performance required of a design solution.
Determining characteristics	Quality Function Deployment Aim: To set targets to be achieved for the engineering characteristics of a product, such that they satisfy customer requirements.
Generating alternatives	Morphological Chart Aim: To generate the complete range of alternative design solutions for a product, and hence to widen the search for potential new solutions.

Stage in the design process	Method relevant to this stage
Evaluating alternatives	Weighted Objectives
	Aim:
	To compare the utility values of alternative design proposals, on the basis of performance against differentially weighted objectives.
Improving details	Value Engineering
	Aim:
	To increase or maintain the value of a product to its purchaser whilst reducing its cost to its producer.

As we shall discuss later, in Chapter 11, these seven stages of design, and their accompanying design methods, should not be assumed to constitute an invariant design process. However, Figure 19 suggests how they relate to each other and to the symmetrical problem-solution model developed in Chapter 2. For example, clarifying objectives (using the Objectives Tree method) is appropriate both for understanding the problem–solution relationship and for developing from the overall problem into sub-problems.

Figure 19
Seven stages of the design process positioned within the symmetrical problem/solution model

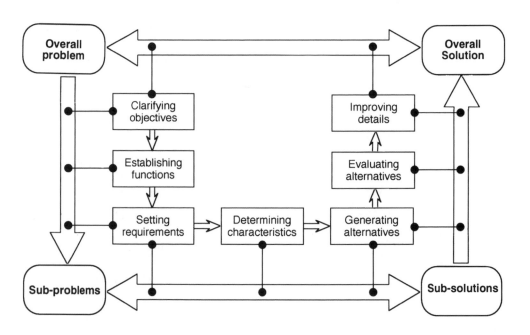

This model of designing integrates the procedural aspects of design with the structural aspects of design problems. The procedural aspects are represented by the sequence of methods (anticlockwise, from top left), and the structural aspects are represented by the arrows showing the commutative relationship between problem and solution and the hierarchical relationships between problem/sub-problems and between sub-solutions/solution.

In the following seven chapters, each of the seven methods included in the model is presented in a step-by-step procedure, followed by a number of short practical examples and a more complete worked example. The examples show that such methods are often adapted to suit the particular requirements of the task in hand. Although it is important not to follow any method in a slavish and unimaginative fashion, it is also important that an effort is made to follow the principles of the method with some rigour. No beneficial results can be expected from slipshod attempts at 'method'.

4 *Clarifying Objectives*

When a client, sponsor or company manager first approaches a designer with a product need, it is unlikely that the 'need' will be expressed very clearly. The client perhaps knows only the type of product that is wanted, and has little idea of the details, or of the variants that might be possible. Or the 'need' might be much vaguer still: simply a 'problem' that needs a solution.

The starting point for a design is therefore very often an ill-defined problem, or a rather vague requirement. It will be quite rare for a designer to be given a complete and clear statement designed object must meet.

An important first step in designing therefore is to try to clarify the design objectives. In fact, it is very helpful at all stages achieve a desired end. That 'end' is the set of objectives that the designed object must meet.

An important first step in designing is therefore to try to clarify the design objectives. In fact, it is very helpful at all stages of designing to have a clear idea of the objectives, even though those objectives may change as the design work progresses. The initial and interim objectives may change, expand or contract, or be completely altered as the problem becomes better understood and as solution ideas develop.

So it is quite likely that both 'ends' and 'means' will change during the design process. But as an aid to controlling and managing the design process it is important to have, at all times, a statement of objectives which is as clear as possible. This statement should be in a form which is easily understood and which can be agreed by the client and the designer, or by the various members of the design team. (It is surprising how often members of the same team can have different objectives!)

The *objectives tree* method offers a clear and useful format for such a statement of objectives. It shows the objectives and the general means for achieving them which are under consideration. It shows in a diagrammatic form the ways in which different objectives are related to each other, and the hierarchical pattern of objectives and sub-objectives. The procedure for arriving at an objectives tree helps to clarify the objectives and to reach agreement between clients, managers and members of the design team.

The Objectives Tree Method

Procedure

Prepare a list of design objectives

The 'brief' for a design problem is often very aptly called that—it is a very brief statement! Such brevity may be because the client is very uncertain about what is wanted, or it may be because he or she assumes that the designer perfectly understands what is wanted. Another alternative is that the client wishes to leave the designer with as much freedom as possible. This might sound like a distinct advantage to the designer, but can lead to great frustration when the client decides that the final design proposal is definitely *not* what was wanted! In any case, the designer will almost certainly need to develop the initial brief into a clear statement of design objectives.

The design objectives might also be called client requirements, user needs or product purpose. Whatever they are called, they are the mixture of abstract and concrete aims that the design must try to satisfy or achieve.

Some design objectives will be contained within the design brief; others must be obtained by questioning the client, or by discussion in the design team. Typically, initial statements of objectives will be brief and rather vague, such as 'The product must be safe and reliable'. To produce more precise objectives, you will need to expand and to clarify such statements.

One way to begin to make a vague statement more specific is, literally, to try to *specify* what it means. Ask 'what is *meant* by that statement?'. For example, an objective for a machine tool that it must be 'safe' might be expanded to mean:

1. Low risk of injury to operator
2. Low risk of operator mistakes
3. Low risk of damage to work-piece or tool
4. Automatic cut-out on overload

This kind of list can be generated simply at random as you think about the objective, or in discussion within the design team. The client may also have to be asked to be more specific about objectives included in the design brief.

The types of questions that are useful in expanding and clarifying objectives are the simple ones of 'Why?' 'How?' and 'What?' For instance, ask 'why do we want to achieve this objective?', 'how can we achieve it?' and 'what implicit objectives underlie the stated ones?' or 'what is the problem really about?'.

Order the list into sets of higher-level and lower-level objectives

As you expand the list of objectives it should become clear that some are at higher levels of importance than others. Sub-objectives for meeting higher-level objectives may also emerge, and some of the statements will be means of achieving certain objectives. This is because some of the questions that you will have been asking about the general objectives imply a 'means–end' relationship—that is, a lower-level objective is a means to achieving a higher-level one.

An example is the statement 'automatic cut-out on overload' in the list above. This is not really an objective in itself, but a means of achieving an objective—in this case, the objective of 'low risk of damage to work-piece or tool'. In turn, this 'low-risk of damage' objective is itself a lower-level objective to that of the overall 'safety' objective.

Your expanded list of objectives will therefore inevitably contain statements at various levels of specificity. In order to clarify the various levels that are emerging, rewrite your general list of objectives into ordered sets. That is, group the objectives into sets, each concerned with one highest-level objective. For example, one set might be to do with 'safety', another to do with 'reliability', and so on. Within each set, list the sub-objectives in hierarchical order, so that the lower-level ones are clearly separated as means of achieving the higher-level ones. Thus, for instance, your 'safety' list might look like this:

Machine must be safe

Low risk of injury to operator
Low risk of operator mistakes
Low risk of damage to work-piece or tool

Automatic cut-out on overload

The list is now ordered into three hierarchical levels. It can sometimes be difficult to differentiate between levels of objectives, or different people in the design team may disagree about relative levels of importance of some objectives. However, exact precision of relative levels is not important, and you want only a few levels, about which most people can agree. For instance, in the above list, 'low risk of injury' might be considered more important than 'low risk of mistakes', but all three 'low risk' objectives can conveniently be grouped at about the same level.

The valuable aspect to sorting objectives roughly into levels is that it encourages you to think more clearly about the objectives, and about the relationships between means and ends. As you write out your lists in hierarchical levels, you will probably also continue to expand them, as you think of further means to meet sub-objectives to meet objectives, etc.

When you have quite a lot of statements of objectives, it is easier to sort them into ordered sets if each statement is written onto a separate slip of paper or small card. Then you can more easily shuffle them about into groups and levels.

Draw a diagrammatic tree of objectives, showing hierarchical relationships and interconnections

As you write out and shuffle your lists, you will probably realize that some of the sub-objectives relate to, or are means of achieving, more than one higher-level objective. For example, the sub-objective of 'low risk of damage to work-piece or tool' might be not only a means of achieving safety but also a means of achieving reliability.

So a diagram of the hierarchical relationships of these few objectives and sub-objectives might look like Figure 20. This diagram is the beginnings of a 'tree' which shows the full pattern of relationships and interconnections. It is not necessarily just a simple 'tree' structure of branches, twigs, leaves, because some of the interconnections form loops or lattices. The 'tree' is also normally drawn 'upside-down', i.e. usually it has increasingly more 'branches' at lower levels, and so it might be better to think of the sub-objectives as 'roots' rather than 'branches'.

It can sometimes be more convenient to draw the 'tree' on its side, i.e. with branches or roots spreading horizontally. In order to help organize the relationships and interconnections between

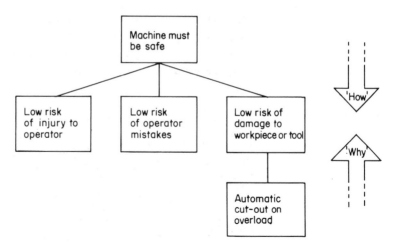

Figure 20
Hierarchical diagram of
relationships

objectives and sub-objectives, draw a complete 'tree' diagram, based on your ordered sets of objectives. Each connecting link that you draw indicates that a lower-level objective is *a means of achieving* the higher-level objective to which it is linked. Therefore working *down* the tree a link indicates *how* a higher-level objective might be achieved; working *up* the tree a link indicates *why* a lower-level objective is included.

Different people might well draw different objectives trees for the same problem, or even from the same set of objectives statements. The tree diagram simply represents one perception of the problem structure. The tree diagram helps to sharpen and improve your own perception of the problem, or to reach consensus about objectives in a team. It is also only a temporary pattern, which will probably change as the design process proceeds.

As with many other design methods, it is not so much the end product of the method (in this case, the tree diagram) which is itself of most value, but the process of working through the method. The objectives tree method forces you to ask questions about objectives, such as 'What does the client mean by X?'. Such questions help to make the design objectives more explicit, and bring them into the open for discussion. Writing the lists and drawing the tree also begins the process of suggesting means of achieving the design objectives, and thus of beginning the process of devising potential design solutions.

Throughout a project, the design objectives should be stated as clearly as the available information permits; the objectives tree facilitates this.

Summary

Aim   To clarify design objectives and sub-objectives, and the relationships between them.

Procedure   1. Prepare a list of design objectives.
These are taken from the design brief, from questions to the client, and from discussion in the design team.
2. Order the list into sets of higher-level and lower-level objectives.
The expanded list of objectives and sub-objectives is grouped roughly into hierarchical levels.
3. Draw a diagrammatic tree of objectives, showing hierarchical relationships and interconnections.
The branches (or roots) in the tree represent relationships which suggest means of achieving objectives.

Examples

Example 1: City transport system

This is an example of expanding and clarifying design objectives from an initially vague brief. A city planning authority asked a transport design team for proposals for 'a modern system, such as a monorail, which would prevent traffic congestion in the city from getting any worse and preferably remove it altogether.'

The only clear objective in this statement is 'To prevent traffic congestion . . . from getting any worse. . .'. But what are the implicit objectives behind the desire for 'a modern system, such as a monorail'? Traffic congestion might be held constant or reduced by other means.

By questioning their clients, the design team uncovered objectives such as a desire to generate prestige for the city and to reflect a progressive image for the city authority. There was also a wish simply to reduce complaints from citizens about the existing traffic system. It was also discovered that only certain types of new system would be eligible for a subsidy from central government.

The design team were able to draw up an expanded and hierarchically-ordered set of objectives, as shown in Figure 21. In particular, they identified a number of high-level, 'essential objectives' which were not explicitly stated in the original brief.

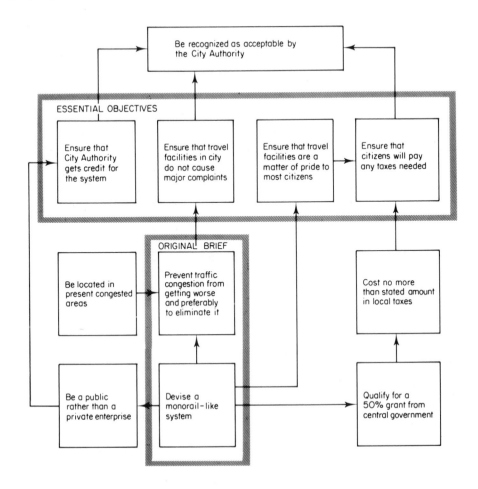

Figure 21
Expanded set of
objectives for a new city
transport system

By identifying these objectives, the designers clarified the project and the limitations that there might be on the range of alternative solutions. (Source: Jones.)

*Example 2: Regional
transport system*

Another example from transport design is shown overleaf, for a larger, regional system. The designers started from the clients' vague definition of 'a convenient, safe, attractive system', and expanded each objective in turn.

For example 'convenience' was defined in terms of 'low journey times' and 'low "out-of-pocket" costs' for users. The latter objective can be met by appropriate pricing policies; low journey times can be met by a variety of sub-objectives, as shown on the left-hand side of the objectives tree in Figure 22.

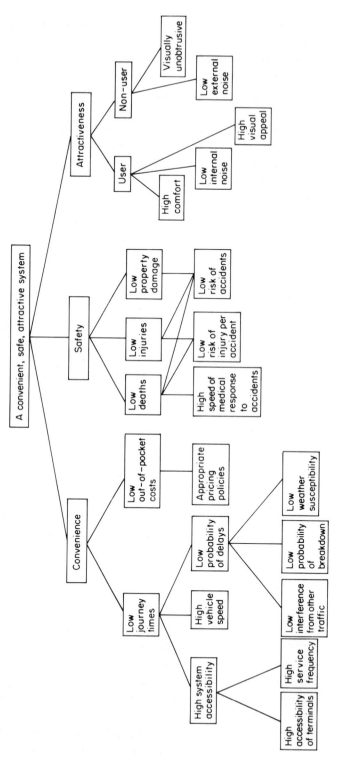

Figure 22   An objectives tree for a 'convenient, safe, attractive' new transport system

Two aspects of 'attractiveness' were defined: user and non-user aspects. The user aspects were subdivided into comfort, visual appeal and internal noise, whereas the non-user aspects were external noise and visual obtrusiveness.

The 'safety' objective was defined to include deaths, injuries and property damage. The sub-objectives to these show how sub-objectives can contribute to more than one higher-level objective. A 'low risk of accidents' can contribute to all three higher-level objectives. If accidents do occur, a 'low risk of injury per accident' can contribute to keeping down both injuries and deaths.

Example 3: Impulse-loading test rig

An example of applying the objectives tree method in engineering design is provided here. The design problem was that of a machine to be used in testing shaft connections subjected to impulse loads.

As before, a typically vague requirement of a 'reliable and simple testing device' can be expanded into a much more detailed set of objectives (Figure 23). 'Reliability' is expanded into 'reliable operation' and 'high safety'. 'Simple' is expanded into 'simple production' and 'good operating characteristics'; the latter is further defined as 'easy maintenance' and 'easy handling'; and so on.

In a case such as this, first attempts at expanding the list of objectives would probably produce statements at all levels of generality. For example, asking 'What is meant by "simple"?' would have been likely to produce statements in random order such as 'Easy maintenance', 'Small number of components', 'Simple assembly', etc. Drawing these out in the hierarchical tree structure shows how they relate together. (Source: Pahl and Beitz.)

Example 4: Automatic teamaker

The objectives tree method can also be used in designing a relatively simple product such as an automatic teamaker. In this example, a distinction is made between 'functions' and 'means'. Each 'function' is an objective, which may be achieved by a number of different 'means' or sub-objectives. Thus the function 'combine water and tea leaves' could be achieved by adding the water to the tea, adding the tea to the water, or bringing them both together into one receptacle (Figure 24).

This is a variation on the objectives tree as described earlier and demonstrated in the other examples, and might more

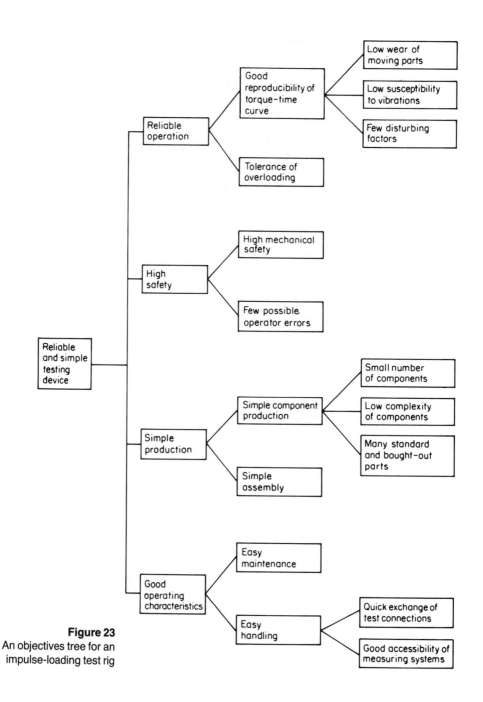

Figure 23
An objectives tree for an
impulse-loading test rig

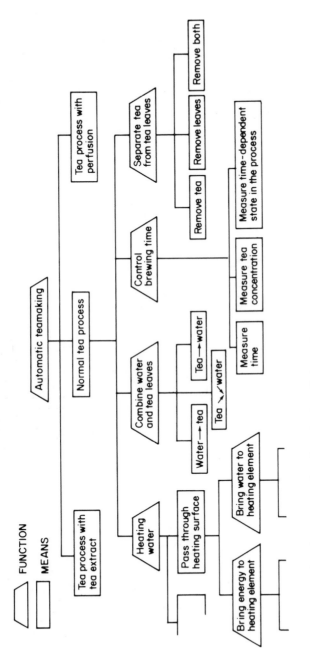

Figure 24   A functions–means tree for an automatic teamaker

accurately be called a 'functions tree'. However, the same principles apply of breaking down objectives into sub-objectives, or functions into means, and ordering them into a hierarchical tree. This application of the tree structure approach helps to ensure that all the possible means of achieving a function (or objective) are considered by the designer. (Source: Tjalve.)

Example 5: Car door

This is another example of a 'function tree'. In considering the requirements for a car door, the designers set out a tree of functions (Figure 25). The tree starts from high-level functional requirements (on the left of the figure), and works through to lower level, detailed functions that can actually be implemented in terms of engineering design decisions. (Source: Pugh.)

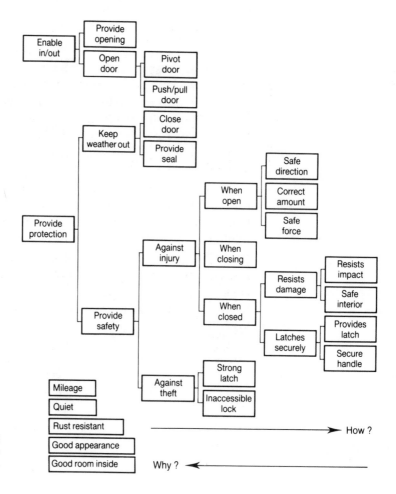

Figure 25
A function tree for a car door

This example is based on the design of a pump for high-pressure, high-temperature fluids. The manufacturers who commissioned the design already made a variety of such pumps, but wished to rationalize their range of pumps in order to reduce manufacturing costs. They also wanted to improve the reliability of their pumps and to offer a product that was seen to be convenient to the varied and changing needs of their customers.

On questioning the client about the 'reliability' and 'convenience' objectives, a common aspect emerged: that the pump should be 'robust'—i.e. that it should not easily fail. The initial list of objectives so far might therefore look like this, in hierarchical order:

> Reliable
> Convenient
> Robust
> Standardized range

These are all still rather high-level and general objectives, so it is necessary to investigate such statements further. In this case, it was possible to investigate the problems experienced with the existing pumps. It was found that they were sometimes affected by cracking and leakages due to the stresses caused by the thermal expansion of the pipes to which they were connected. This appeared to be the main problem to which the requests for 'robustness' and 'reliability' were aimed.

Similarly, investigating the 'convenient' objective revealed a further two sub-objectives; firstly that the pumps should be easy to install and replace, and secondly that they should occupy the minimum space. It was realized that the standardization of sizes and dimensions in the range could be a means of helping to achieve these objectives, as well as reducing manufacturing costs.

The expanded list of objectives therefore looked like this:

Reliable

Robust	Convenient
Resistant to external mechanical stresses	Easy to install and replace
Unaffected by thermal expansion of pipes	Occupy minimum space
	Standardized range

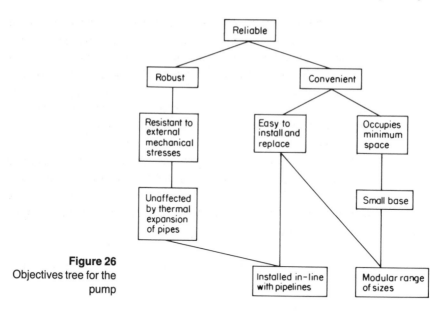

Figure 26
Objectives tree for the
pump

Figure 27   The Danish Grundfos pumps which have been designed on similar principles to the objectives developed in this example

A key design principle to emerge from considering the means of achieving these objectives was that the inlet and outlet ports should always be in line, to avoid the thermal expansion problems. Such a system, coupled with a small base size and modular dimensioning of alternative components, would also facilitate installation and replacement of the pump. The objectives tree therefore looked like Figure 26.

A high-pressure pump has been designed on similar principles in Denmark (see Figure 27). According to the Danish Design Council, the pump is 'almost a diagram of its problem statement: intake and discharge are aligned, motor, coupling and the stage-built pump are aligned on an axis at right angles to the installation surface, and the pump pressure is increased by adding to the number of stages, i.e. a change in height. The pump is installed directly on the pipeline, occupying a minimum of space.'

5 *Establishing Functions*

We have seen from the objectives tree method that design problems can have many different levels of generality or detail. Obviously, the level at which the problem is defined for or by the designer is crucial. There is a big difference between being asked to 'design a telephone handset' and to 'design a telecommunication system'.

It is always possible to move up or down the levels of generality in a design problem. The classic case is that of the problem 'to design a doorknob'. The designer can move up several levels to that of designing the door or even to designing 'a means of ingress and egress' and find solutions which need no doorknob at all—but this is of no use to a client who manufactures doorknobs! Alternatively, the designer can move down several levels, investigating the ergonomics of handles or the kinematics of latch mechanisms—perhaps again producing non-doorknob solutions which are functional improvements but which are not what the client wanted.

However, there are often occasions when it is appropriate to question the level at which a design problem is posed. A client may be focusing too narrowly on a certain level of problem definition, when a resolution at another level might be better, and reconsidering the level of problem definition is often a stimulus to the designer to propose more radical or innovative types of solutions.

So it is useful to have a means of considering the problem level at which a designer or design team is to work. It is also very useful if this can be done in a way that considers, not the potential type of solution, but the essential functions that a solution type will be required to satisfy. This leaves the designer free to develop alternative solution proposals that satisfy the functional requirements.

The *function analysis* method offers such a means of considering essential functions and the level at which the problem is to be addressed. The essential functions are those that the device, product or system to be designed must satisfy, no matter what physical components might be used. The problem level is decided by establishing a 'boundary' around a coherent sub-set of functions.

The Function Analysis Method

Procedure

Express the overall function for the design in terms of the conversion of inputs into outputs

The starting point for this method is to concentrate on what has to be achieved by a new design, and not on how it is to be achieved. The simplest and most basic way of expressing this is to represent the product or device to be designed as simply a 'black box' which converts certain 'inputs' into desired 'outputs'. The 'black box' contains all the functions which are necessary for converting the inputs into the outputs (Figure 28).

Figure 28
The 'black box' systems model

It is preferable to try to make this overall function as broad as possible at first—it can be narrowed down later if necessary. It would be wrong to start with an unnecessarily limited overall function which restricts the range of possible solutions. The designer can make a distinct contribution to this stage of the design process by asking the clients or users for definitions of the fundamental purpose of the product or device, and asking about the required inputs and outputs—from where do the inputs come?, what are the outputs for?, what is the next stage of conversion?, etc.

This kind of questioning is known as 'widening the system boundary'. The 'system boundary' is the conceptual boundary that is used to define the function of the product or device. Often, this boundary is defined too narrowly, with the result that only minor design changes can be made, rather than a radical rethinking.

It is important to try to ensure that all the relevant inputs and outputs are listed. They can all usually be classified as flows of either materials, energy or information, and these same classifications can be used to check if any input or output type has been omitted.

Break down the overall function into a set of essential sub-functions

Usually, the conversion of the set of inputs into the set of outputs is a complex task inside the 'black box', which has to be broken down into sub-tasks or sub-functions. There is no really objective, systematic way of doing this; the analysis into sub-functions may depend on factors such as the kinds of components available for specific tasks, the necessary or preferred allocations of functions to machines or to human operators, the designer's experience, and so on.

In specifying sub-functions it is helpful to ensure that they are all expressed in the same way. Each one should be a statement of a verb plus a noun—for example, 'amplify signal', 'count items', 'separate waste', 'reduce volume'.

Each sub-function has its own input(s) and output(s), and compatibility between these should be checked. There may be 'auxiliary sub-functions' that have to be added but which do not contribute directly to the overall function, such as 'remove waste'.

Draw a block diagram showing the interactions between sub-functions

A block diagram consists of all the sub-functions separately identified by enclosing them in boxes and linked together by their inputs and outputs so as to satisfy the overall function of the product or device that is being designed. In other words, the original 'black box' of the overall function is redrawn as a 'transparent box' in which the necessary sub-functions and their links can be seen (Figure 29).

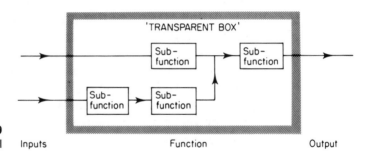

Figure 29
A 'transparent box' model

In drawing this diagram you are deciding how the internal inputs and outputs of the sub-functions are linked together so as to make a feasible, working system. You may find that you have to juggle inputs and outputs, and perhaps redefine some sub-functions so that everything is connected together. It is useful to use different conventions, i.e. different types of lines, to show the different types of inputs and outputs, i.e. flows of materials, energy or information.

Draw the system boundary

In drawing the block diagram you will also need to make decisions about the precise extent and location of the system boundary. For example, there can be no 'loose' inputs or outputs in the diagram except those that come from or go outside the system boundary.

It may be that the boundary now has to be narrowed again, after its earlier broadening during consideration of inputs, outputs and overall function. The boundary has to be drawn around a sub-set of the functions that have been identified, in order to define a feasible product. It is also probable that this drawing of the system boundary is not something in which the designer has complete freedom—as likely as not, it will be a matter of management policy or client requirements. Usually, many different system boundaries can be drawn, defining different products or solution types.

Search for appropriate components for performing the sub-functions and their interactions

If the sub-functions have been defined adequately and at an appropriate level, then it should be possible to identify a suitable component for each sub-function. This identification of components will depend on the nature of the product or device, or more general system, that is being designed. For instance, a 'component' might be defined as a person who performs a certain task, a mechanical component, or an electronic device. One of the interesting design possibilities opened up by electronic devices such as micro-processors is that these can often now be substituted for components that were previously mechanical devices or perhaps could only be done by human operators. The Function Analysis method is a useful aid in these circumstances because it focuses on functions, and leaves the physical means of achieving those functions to this later stage of the design process.

Summary

Aim

To establish the functions required, and the system boundary, of a new design.

Procedure

1. Express the overall function for the design in terms of the conversion of inputs into outputs.
The overall, 'black box' function should be broad—widening the system boundary.

2. Break down the overall function into a set of essential sub-functions.
The sub-functions comprise all the tasks that have to be performed inside the 'black box'.

3. Draw a block diagram showing the interactions between sub-functions.
The 'black box' is made 'transparent', so that the sub-functions and their interconnections are clarified.

4. Draw the system boundary.
The system boundary defines the functional limits for the product or device to be designed.

5. Search for appropriate components for performing the sub-functions and their interactions.
Many alternative components may be capable of performing the identified functions.

Examples

Example 1: A feed delivery system

The function analysis method is particularly relevant in the design of flow-process systems, such as that shown diagrammatically in Figure 30. This represents a factory where animal feedstuffs are bagged.

In this example, the company wanted to try to reduce the relatively high costs of handling and storing the feedstuffs. A designer might tackle this task by searching for very direct ways in which each part of the existing process might be made more cost-effective. However, a broader formulation of the problem—the overall function—was represented in the following stages:

1	Stacked sacks await filling	2	Man C folds and stitches the top of the bag
1	Man A lifts empty sack from stack and places it under spout for filling	4	Man D takes the bag and loads it on wagon
1	Man A fills the 100-pound sack by gravity feed, manually controlling the rate of flow	5	Loaded wagon is pushed to warehouse
2	Man A hands the bag to man B	6	Bags are stacked by men E and F
1	Man B checks the weight and adds or removes material when necessary to adjust the weight to approximately 100 pounds	7	Bags are stored awaiting sale
3	Man B hands the bag to man C	7	Bags are loaded on waiting truck, two or three at a time by handtruck, then delivered to consumer

Figure 30   The existing method of filling, storing and dispatching bags of animal feed

1. Transfer of feed from mixing bin to bags stored in warehouse.
2. Transfer of feed from mixing bin to bags loaded on truck.
3. Transfer of feed from mixing bin to consumers' storage bins.
4. Transfer of feed ingredients from source to consumers' storage bins.

This broadening of the problem formulation is shown diagrammatically in Figure 31.

Each different formulation suggests different kinds of solutions, with the broadest formulation perhaps leading to the complete elimination of the handling, storing and loading sub-functions. (Source: Krick.)

Example 2: Packing carpet squares

This example shows another flow process—the packing of loose carpet squares into lots. The designers first broke down the overall function into a series of principal sub-functions (Figure 32).

Some auxiliary functions then became clear. For example, the input from the separate stamping machine includes off-cuts which have to be removed; reject squares must also be removed; materials must be brought in for packaging. The sub-function 'count squares' could also be used to give the signal for packaging lots of a specified number (see Figure 33). (Source: Pahl and Beitz.)

Figure 31
Alternative formulations
of the feed distribution
problem

Figure 32
Analysis of principal
functions for the packing
of carpet squares

Figure 33
Expanded function
analysis for the packing
of carpet squares

*Example 3: Automatic
teamaker*

This example is a further development of the project for the
design of an automatic teamaking machine, started in the objectives tree method (Figure 24). The fundamental process to be
achieved by such a machine is to convert cold water and tea
leaves into hot tea (there will also be a need to remove waste
tea leaves after the brewing process). This overall function is
shown in 'black box' form in Figure 34.

Figure 34
Black box model of the
tea brewing process

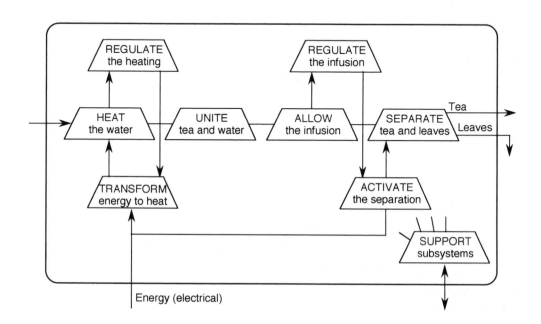

Figure 35
Three alternative process
models for tea brewing

Figure 36
Function analysis for the
automatic teamaker

After considering various alternative processes by which the overall function can be achieved, the designer settled on the process shown as a flow diagram of sub-functions in Figure 35(a). Various necessary auxiliary functions then became apparent, particularly to do with controlling the heating and brewing processes. The resulting function analysis diagram is shown in Figure 36. (Source: Hubka *et al.*)

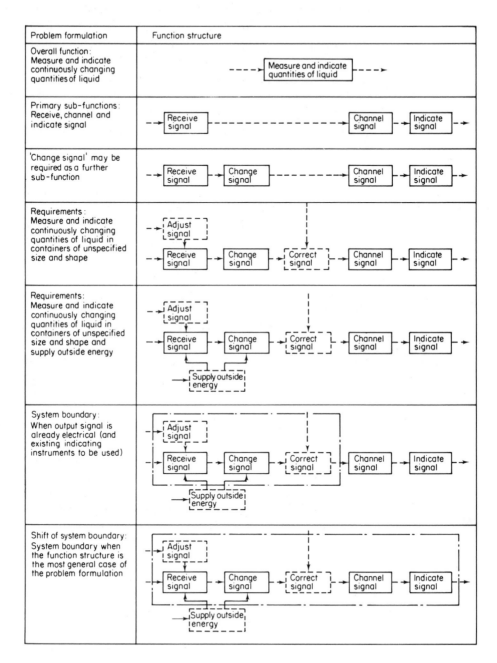

Figure 37 Function analysis for a fuel gauge

Example 4: Fuel gauge

Function analysis can also be applied in the design of much smaller products or devices. Figure 37 shows the step-by-step development of a function analysis for a fuel gauge. Notice how auxiliary functions are introduced in order to cope with a gradually broadening problem formulation to provide for fuel containers of different sizes and shapes, etc. The Figure also shows how the system boundary can be drawn in different places, in this case depending on whether the output signal is to be to already-existing instruments or whether such an instrument is to be included as an integral part of the design. (Source: Pahl and Beitz.)

Worked example: Washing machine

A relatively simple example of the use of the function analysis method is based on the domestic washing machine. The overall function of such a machine is to convert an input of soiled clothes into an output of clean clothes, as shown in Figure 38.

Figure 38
Black box model of a
washing machine

Inside the 'black box' there must be a process that separates the dirt from the clothes, and so the dirt itself must also be a separate output. We know that the conventional process involves water as a means of achieving this separation, and that a further stage must therefore be the conversion of clean (wet) clothes to clean (dry) clothes. Even further stages involve pressing and sorting clothes. The inputs and outputs might therefore be listed like this:

Inputs	*Outputs*	
Soiled clothes	(Stage 1)	Clean clothes
		Dirt
	(Stage 2)	Dry clothes
		Water
	(Stage 3)	Pressed clothes

The essential sub-functions, together with the conventional means of achieving them, for converting soiled clothes into

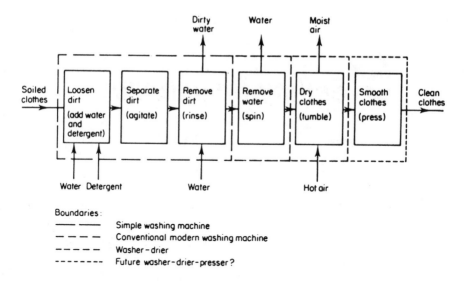

Dirty   Water   Moist
water   air

Soiled | Loosen | Separate | Remove | Remove | Dry | Smooth | Clean
clothes | dirt | dirt | dirt | water | clothes | clothes | clothes
(add water	(agitate)	(rinse)	(spin)	(tumble)	(press)
and					
detergent)					

Water Detergent   Water   Hot air

Boundaries:
——— —— Simple washing machine
— — — — Conventional modern washing machine
– – – – – Washer–drier
- - - - - - - Future washer–drier–presser?

Figure 39
Function analysis of the
washing machine

clean, pressed clothes would therefore be as follows:

Essential sub-functions	*Means of achieving sub-functions*
Loosen dirt	Add water and detergent
Separate dirt from clothes	Agitate
Remove dirt	Rinse
Remove water	Spin
Dry clothes	Blow with hot air
Smooth clothes	Press

A block diagram, with main and subsidiary inputs and outputs, might look like Figure 39. The development of washing machines has involved progressively widening the system boundary, as shown in the figure. Early washing machines simply separated the dirt from the clothes, but did nothing about removing the excess water from the clothes; this was left as a task for the human operator, using either a hand or mechanical wringing of the clothes. The inclusion of a spin-drying function removed the excess water, but still left a drying process. This is now incorporated in washer-driers. Perhaps the smoothing of clothes will somehow be incorporated in future machines? (Although this need has been reduced by the use of artificial fabrics in clothes.)

6 *Setting Requirements*

Design problems are always set within certain limits. One of the most important limits, for example, is that of cost: what the client is prepared to spend on a new machine, or what customers may be expected to pay as the purchase price of a product. Other common limits may be the acceptable size or weight of a machine; some limits will be performance requirements, such as an engine's power rating; still others might be set by statutory legal or safety requirements.

This set of requirements comprises the performance specification of the product or machine. Statements of design objectives or functions (such as those derived from objectives tree or function analysis methods) are sometimes regarded as being performance specifications, but this is not really correct. Objectives and functions are statements of what a design must achieve or do, but they are not normally set in terms of precise limits, which is what a performance specification does.

In setting limits to what has to be achieved by a design, a performance specification thereby limits the range of acceptable solutions. Because it therefore sets the designer's target range, it should not be defined too narrowly. If it is, then a lot of otherwise acceptable solutions might be eliminated unnecessarily. On the other hand, a specification that is too broad or vague can leave the designer with little idea of the appropriate direction in which to aim. Specification limits that are set too wide can also lead to inappropriate solutions which then have to be changed or modified when it is found that they actually fall outside acceptable limits.

So there are good reasons for putting some effort into an accurate performance specification early in the design process. Initially, it sets up some boundaries to the 'solution space' within which the designer must search. Later on in the design process, the performance specification can be used in evaluating proposed solutions, to check that they do fall within the acceptable boundaries.

The *performance specification* method is intended to help in defining the design problem, leaving the appropriate amount of freedom so that the designer has room to manoeuvre over the ways and means of achieving a satisfactory design solution. A specification defines the required *performance*, and not the required *product*. The method therefore emphasizes the performance that a design solution has to achieve, and not any particular physical components which may be means of achieving that performance.

The Performance Specification Method

Procedure

Consider the different levels of generality of solution which might be applicable

It is important that a specification is addressed to an appropriate level of generality for the solution type that is to be considered. A specification at too high a level of generality may allow inappropriate solutions to be suggested, whereas too low a level (a specification which is too specific) can remove almost all of the designer's freedom to generate a range of acceptable solutions.

So the first step is to consider the different levels of generality. A simple classification of types of level, from the most general down to the least, for a product might be:

product *alternatives*
product *types*
product *features*

As an example to illustrate these levels, suppose that the product in question is a domestic heating appliance. At the highest level of generality the designer would be free to propose alternative ways of heating a house, such as moveable appliances, fixed appliances, central heating with radiators, ducted warm air, etc. There might even be freedom to move away from the concept of an 'appliance' to alternative forms of heating such as conservatories that trap solar heat; or to ways of retaining heat, such as insulation. At the intermediate level, the designer would have a much more limited freedom, and might only be concerned with different types of appliance, say, different heater types such as radiators or convectors, or different fuel types. At the lowest level, the designer would be constrained to considering different features within a particular type of appliance, such as its heating element, switches, body casing, supports, etc.

Determine the level of generality at which to operate

Considering the different levels of generality might lead to either a broadening or a narrowing of initial product concepts or of the design brief. The second step of the method is therefore to make a decision on the appropriate level.

Normally, the client, company management or customer decides the level at which the designer will operate. For instance, in the case of domestic heating appliances, the highest level of generality ('alternatives') would only be considered if an appliance manufacturer was proposing to diversify or broaden its activities into other aspects of domestic heating. The intermediate level ('types') would normally be considered when a new product was to be designed, to add to the existing range of appliances or to replace obsolete ones. The lowest level ('features') would be considered when making modifications to existing products.

The higher the level of generality that may be considered, the more freedom the designer has in terms of the range of acceptable solutions. Of course, the higher levels also subsume the lower levels of specification—that is, the specification of features is part of the specification of types which is part of the specification of alternatives.

Identify the required performance attributes

Once the level at which designing is to proceed has been decided, work can begin on the performance specification proper. Any product or machine will have a set of *attributes*, and it is these which are specified in the performance specification. Attributes include such things as comfort, portability and durability, and key features such as speed, cost and safety.

Performance attributes are usually similar to, or derived from, the design objectives and functions. So if you have already prepared an objectives tree or a function analysis, these are likely to be the source of your initial list of performance attributes.

A most important aspect to bear in mind when listing performance attributes is that they should be stated in a way which is independent of any particular solution. Statements of attributes made by clients or customers are often couched in terms of solutions, because they value some performance aspect which is embodied in the solution but they have not separated the attribute from a particular embodiment. Such solution-based, rather than performance-based, statements are usually unnecessarily restrictive of solution concepts.

For example, a client might suggest that the material for a particular surface area should be ceramic tiles, because that is a satisfactory feature of an existing solution. But the essential performance requirement might be that the surface should be non-porous, or easy to clean, or have a smooth, hard texture, or simply have a shiny appearance. Acceptable alternatives might be plastics, metal or marble.

There may be a whole complex of reasons underlying a client or customer specification of a particular solution feature. It could be the whole set of attributes of a ceramic surface, as just listed, plus the mass which is provided by ceramic tiles, plus the colour range, plus some perceived status or other value which is not immediately obvious. A comprehensive and reliable list of performance attributes can therefore take some considerable effort to compile, and may well require careful research into client, customer, and perhaps manufacturer requirements.

The final list of performance attributes contains all the conditions that a design proposal should satisfy. However, it may become necessary to distinguish within this list between those attributes or requirements that are 'demands' and those that are 'wishes'. 'Demands' are requirements that must be met, whereas 'wishes' are those that the client, customer or designer would like to meet if possible. For example, the requirement of a non-porous surface might be a functional 'demand', but availability in a range of colours might be a 'wish' dependent on the material actually chosen.

State succinct and precise performance requirements for each attribute

Once a reliable list of attributes has been compiled, a performance specification is written for each one. A specification says what a product must *do*, not what it must *be*. Again, this may well require some careful research—it is not adequate simply to guess at performance requirements, nor just to take them from an existing solution type.

Wherever possible, a performance specification should be expressed in quantified terms. Thus, for example, a maximum weight should be specified, rather than a vague statement such as 'lightweight'. A safety specification—say, for escape from a vehicle—should state the maximum time allowable for escape in an emergency, rather than using terms like 'rapidly' or 'readily'.

Also wherever possible and appropriate, a specification should set a range of limits within which acceptable performance lies. So a specification should not say 'Seat height: 425mm' if a range between 400mm and 450mm is acceptable. On the other hand, spurious 'precision' is also to be avoided: don't specify 'A container of volume 21.2 l' if you mean to refer to a waste-paper bin 'Approximately 300mm diameter and 300mm high'.

Summary

Aim

To make an accurate specification of the performance required of a design solution.

Procedure

1. Consider the different levels of generality of solution which might be applicable.
There might be a choice between

   product alternatives
   product types
   product features

2. Determine the level of generality at which to operate.
This decision is usually made by the client. The higher the level of generality, the more freedom the designer has.

3. Identify the required performance attributes.
Attributes should be stated in terms which are independent of any particular solution.

4. State succinct and precise performance requirements for each attribute.
Wherever possible, specifications should be in quantified terms, and identify ranges between limits.

Examples

This example is a specification for a domestic water mixing tap that can be operated with one hand (Figure 40). The initial design brief for this project was given in Chapter 1 (page 11). Notice how the brief has been considerably expanded, as the design team has researched the problem. Some details in the brief have changed (for example, the maximum pressure) as a result of establishing the national standards that apply to such a product. The range of users has also been taken into account (requirement 11—light operation for children), as have safety considerations (requirements 18–20). The project time-scale has also been included in the specification. The 'D or W' column on the left distinguishes between 'Demands' and 'Wishes' in the specification. (Source: Pahl and Beitz.)

This problem was formulated by the client at the lowest level of generality: the design of a particular type of fuel gauge for use in motor vehicles. The initial general formulation of the problem statement was:

> A gauge to measure continuously changing quantities of liquid in containers of unspecified size and shape, and to indicate the measurement at various distances from the containers.

The following list of attributes was then developed:

Suitable for containers (fuel tanks) of
  various volumes
  various shapes
  various heights
  various materials
Connection to top or side of container
Operates at various distances from container
Measures petrol or diesel liquid
Accurate signal
Reliable operation

The design team went on to develop a full performance specification, as shown in Figure 41. As in the previous example, they also distinguished between 'Demands' (D) and 'Wishes' (W). (Source: Pahl and Beitz.)

		Specification		
		for One-handed mixing tap		Page 1
Changes	D or W	Requirements		Responsible
	D	1 Throughput (mixed flow) max. 10 l/min at 2 bar		
	D	2 Max. pressure 10 bar (test pressure 15 bar as per DIN 2401)		
	D	3 Temp. of water: standard 60°C, 100°C (short-time)		
	D	4 Temperature setting independent of throughput and pressure		
	W	5 Permissible temp. fluctuation ±5°C at a pressure diff. of ±5 bar between hot and cold supply		
	D	6 Connection: 2 × Cu pipes, 10 × 1 mm, l = 400 mm		
	D	7 Single-hole attachment $\phi\ 35^{+2}_{-1}$ mm, basin thickness 0 – 18 mm (Observe basin dimensions DIN EN 31, DIN EN 32, DIN 1368)		
	D	8 Outflow above upper edge of basin: 50 mm		
	D	9 To fit household basin		
	W	10 Convertible into wall fitting		
	D	11 Light operation (children)		
	D	12 No external energy		
	D	13 Hard water supply (drinking water)		
	D	14 Clear identification of temperature setting		
	D	15 Trade mark prominently displayed		
	D	16 No connection of the two supplies when valve shut		
	W	17 No connection when water drawn off		
	D	18 Handle not to heat above 35°C		
	W	19 No burns from touching the fittings		
	W	20 Provide scalding protection if extra costs small		
	D	21 Obvious operation, simple and convenient handling		
	D	22 Smooth, easily cleaned contours, no sharp edges		
	D	23 Noiseless operation (⩽20 dB as per DIN 52218)		
	W	24 Service life 10 years at about 300 000 operations		
	D	25 Easy maintenance and simple repairs. Use standard spare parts		
	D	26 Max. manuf. costs DM 30 (3000 units per month)		
	D	27 Schedules from inception of development		

	Conceptual design	Embodiment design	Detail Design	Prototype	
after	2	4	6	9	months

Figure 40   Specification for a one-handed mixing tap

Figure 41 Specification for the fuel gauge

	D or W	*Specification*	
		for Fuel gauge	*Page 2*

Changes	D or W	Requirements	Responsible
	W	Possibility of signal calibration with full container	
	D	Minimum measurable content: 3% of maximum value	
	W	Reserve tank contents by special signal	
		4. Operating conditions	
	D	Forward acceleration ± 10 m/s^2	
	D	Sideways acceleration ± 10 m/s^2	
	D	Upward acceleration (vibration) up to 30 m/s^2	
	W	Shocks in forward direction without damage up to 30 m/s^2	
	D	Forward tilt up to $\pm 30°$	
	D	Sideways tilt max. 45°	
	D	Tank not pressurized (ventilated)	
		5. Test requirements	
	D	Salt spray tests for inside and outside components according to client's requirements	
	D	Pressure test for container 30 kN/m^2	
		6. Life expectancy, durability of container	
	D	Life expectancy 5 years in respect of corrosion due to contents and condensation	
	D	Must conform with heavy vehicle requirements	
		7. Production	
	W	Simply modified to suit different container sizes	
		8. Operation, maintenance	
	W	Installation by non-specialist	
	D	Must be replaceable and maintenance-free	
		9. Quantity	
		10 000/day of the adjustable type, 5000/day of the most popular type	
		10. Costs	
		Manufacturing costs $\leqslant$ DM 3.00 each	

This example shows the development of a performance specification for a consumer product: an electric toothbrush. The problem is set at the intermediate level of generality, i.e. a new type of toothbrush, but it has novel features which require precise performance specifications.

The designers listed the new product's attributes mainly in terms of a set of 'user needs':

User needs at the concept phase

Physiological needs	Clean teeth better than a handbrush, massage gums, reduce decay, hygienic family sharing, electrical and mechanical safety, etc.
Social needs	Sweet breath and white teeth (symbolic needs for social acceptance); handle colours to match bathroom, etc.
Psychological needs	Autonomy in deciding when and how one's teeth are to be cared for, self-esteem from care of teeth, praise for effort, pleasure from giving or receiving a gift, etc.
Technical needs	Diameter, length, brush size, amplitude, frequency, weight, running time, reliability, useful life, etc.
Time needs	Needed for Christmas market
Resources exchanged	$1 per person is the lowest cost alternative, but electric razors sell for 20 times the price of a manual razor, so probably $20 will be paid for an electric toothbrush

The performance specification was then drawn up as a set of design objectives with corresponding criteria, as shown in Figure 42. (Source: Love.)

Objectives	Criteria
1. To be attractive, suitable for sale primarily in the gift market and secondly as a personal purchase.	1a. Attractiveness of overall design and packaging to be judged better than brands X and Y by more than 75% of a representative consumer panel. 1b. Decorator colours to be the same as our regular products. 1c. Package can be displayed on counter area of 75 × 100 mm.
2. The technical functions are to be at least as good as past 'family' models of brand X.	2a. Technical functions to be judged at least as good as the past 'family' model of brand X by dental consultant, Dr J.P. 2b. Amplitude to be between 2 and 3 mm. 2c. Frequency to be 15 ± 5 cycles/s. 2d. Battery life to be minimum of 50 min. when tested according to standard XYZ. 2e. etc., for other technical aspects such as weight, impact strength, frequency of repair, dimensions . . . .
3. To be saleable in the United States and Canada.	3. Must meet UL and CSA standards for safety (a crucial criterion).
4. The timing objective is that the product be ready for sale to the Christmas trade in the nearest feasible season.	4. The time milestones, backing up from October production are to be: • mock-up approval—2 months • tooling release—6 months • production prototype—10 months • pilot run—10 months • production run—13 months (October).
5. The selling price is to be not more than 10% of the present utility models.	5. The selling price is to be between $12.50 and $17.50, depending on the features offered, for a production run of 100 000 units.

Figure 42   Performance specification for an electric toothbrush

Example 4: City car   In this example, a design team undertook a study of the 'city car' concept, i.e. a small runabout for use in cities or other limited-journey purposes. Many different solutions have been developed to this problem, with varying degrees of success.

The design study included an analysis of the features of the many previous examples of city car designs, as well as market research, town planning and engineering criteria, etc. As a result, a specification of the general features required or desired in a city car was drawn up, and is shown in Figure 43. The characteristics are classified as either 'required' or 'desired'—

	Characteristics	
	Required	Desired
1 General characteristics		
Car for city or for delimited area (airport, port, railway station, industry, etc.)	X	
Number of seats: 2(4)	X	
Number of wheels: 4	X	
Utilization of space: maximum in relation to the external dimensions		X
Vehicle must be economical	X	
Range > 100 km		X
2 Working conditions		
The same as those relative to the use of every automobile in city or for delimited area	X	
3 Dimensions		
Length 2.5 m	X	
Width 1.5 m		X
Height 1.6 m		X
4 Weight		
Maximum net weight 400 kg		X
Available loading capacity 200 kg (300 kg)	X	
Gross weight 600 kg (700 kg)		X
5 Capacity for luggage		
Minimum volume 150 dm^3	X	
With back row seats dropped down 350 dm^3		X
Possibility for loading outside (on the roof)		X
6 Velocity		
Average velocity 50 km/h	X	
Maximum velocity 70 km/h	X	
7 Type of motor		
Thermal (internal combusion, etc.)	X	
Electrical		
Thermoelectrical		
Thermohydraulic		
Air compressed		
8 Characteristics of motor		
Power: 2 kW		X
Fuel consumption 1 litre per 25 km		X
(thermic motor)		
9 Maintenance		
Minimum and simple		X
Possibility of finding spare parts: easy		X
10 Utilization, use		
Utilization: simple		X
Use: frequent	X	
Reliability: high	X	
11 Durability		
At least 50 000 km or 5 years	X	
12 Safety		
As high as possible (active and passive)	X	
13 Pollution		
Not above the legal minimum	X	
Zero		X
14 Form and aesthetics		
Pleasant and therefore able to be commercialized	X	
Convertible		X
15 Production		
In small series (500 vehicles/year)		X
16 Price		
$2000 to $2500 (2/3 Fiat 126)	X	

Figure 43   Specification for a small city-car

similar to 'demands' and 'wishes' in other examples. (Source: Pighini *et al.*)

Worked example:
Portable computer

Personal computers have become a fundamental aspect of many activities. Some people need to have not only one at the office and one at home, but also one that can be used at other locations and that therefore travels with them. This example is based on the design of such a portable computer.

Clearly, the level of generality for this problem has been set by the client's request for a new design of portable computer: it is a particular product type, with the designers' freedom therefore constrained to product *features*.

There are many specialized attributes which would have to be researched and specified, such as the type of microprocessors to be incorporated, type of screen display, keyboard, etc. We shall concentrate here only on the key attribute of 'portable'. What exactly does this mean? We need to know what features of 'portability' might be important to potential purchasers of the computer.

We therefore interview a range of computer users about their needs. From this it emerges that there are two distinct aspects to 'portability'. The first is, quite simply, that the machine can be carried comfortably and easily—in contrast to some early models, which were known as 'luggable' rather than 'portable'! The second aspect is that the purpose of a portable machine is that it can be used in a wide variety of different locations.

Further research with users is necessary to develop performance specifications for both of these attributes. For example, to specify the 'carryable' performance attribute, it is not adequate simply to suggest a carrying handle. Nor is it adequate just to weigh a rival product and specify that as a maximum weight. We need to know the range of users for the computer and the typical distances or lengths of time that it might be carried. Experiments with a few representative, least-strong users and maximum expected journeys could then establish an appropriate weight limit.

We also need to investigate further the variety of locations in which it is desired to use the computer. Does this include on someone's lap on a train or plane? In that case the machine must be small but stable. Does it include word-processing use during meetings, conferences or lectures? In that case its operation should be silent or very quiet. Does it include outdoor use? In that case there might be weatherproofing requirements, or the user might be wearing gloves.

Obviously, in many locations there is no available power source, and so a portable computer must have its own batteries. But it may be that salespeople and others often use the computer in their car, perhaps in conjunction with a fax machine and telephone, and therefore the car's battery could be used through the cigarette-lighter socket. One other aspect of performance that emerges is that users want to be able to plug the portable computer into conventional devices such as VDUs and printers at different locations in order to provide a display visible by a group of people or to leave a hard-copy printout. This means that hardware compatibility with such devices is important.

A performance specification for the 'portability' attribute is therefore developed as follows:

Performance specification Portable computer
Portability
    Can be carried in one hand
    Weight not more than 5 kg
    Optional carrying case, with pockets for disks
    When closed, weatherproof against rain showers
    Maximum base dimensions in use: 450 mm × 450 mm
    Near-silent keyboard
    Power sources:
      mains
      own battery
      car cigarette-lighter socket
    Compatible with standard VDUs and printers

7 *Determining Characteristics*

In determining a product specification, conflict and misunderstanding can sometimes arise between the marketing and the engineering members of the design team. This is usually because they focus on different interpretations of what should be specified. Managers and market researchers tend to concentrate more on specifying the desirable *attributes* of a new product (usually from the viewpoint of customer or client requirements), whereas designers and engineers concentrate more on a product's engineering *characteristics* (usually in terms of its physical properties).

The relationship between characteristics and attributes is in fact a very close one, and confusion can be avoided if this relationship is clearly understood. Designers make decisions about the product's physical properties, and thus determine its engineering characteristics; but those characteristics then determine the product's attributes, which in turn satisfy the customer's needs and requirements. Thus, for example, the engineering designer may choose a particular metal casing for a product, of a certain gauge and surface finish, thus determining characteristics such as weight, rigidity and texture; these characteristics determine product attributes such as portability, durability and appearance.

With increased competition in all product markets, it has become necessary to ensure that this relationship between engineering characteristics and product attributes is properly understood. In particular, it is necessary to understand just what customers want in terms of product attributes and to ensure that these are carefully translated into specifications of the appropriate engineering characteristics. This attitude towards product design is based on the philosophy of 'listening to the voice of the customer', and is reflected in an increased concentration on product quality. Design for quality is recognized as a major factor in determining the commercial success of a product.

A comprehensive method for matching customer requirements to engineering characteristics is the *quality function deployment* method (QFD). 'Quality function deployment' is a direct translation of the Japanese characters *Hin Shitsu, Ki No, Ten Kai*. In Japanese, the phrase means something like the strategic arrangement (deployment) throughout all aspects of a product (functions) of appropriate characteristics (qualities) according to customer demands.

The QFD method recognizes that the person who buys (or who most influences the buying decision for) a product is the most important person in determining the commercial success of a product. If customers do not buy it, then the product—however 'well-designed' it may be—will be a commercial failure. Therefore 'the voice of the customer' has priority in determining the product's attributes. This means taking care to identify who the customers are, to listen carefully to what they say, and to determine the product's engineering characteristics in the light of this.

QFD is essentially concerned with the translation of customer requirements into engineering characteristics, and it is presented here at the core of the design process. However, because it is a comprehensive method, aspects of QFD can be used at various stages of the design process, and it also draws upon features from several other design methods.

The Quality Function Deployment Method

Procedure

Identify customer requirements in terms of product attributes

The method starts with the identification of the customers and of their own views of their requirements and desired product attributes. There are various market research techniques that can be used to assist the gathering of information about customer requirements and preferences. These methods include product 'clinics' where customers are quizzed in depth about what they like and dislike about particular products, and 'hall tests' where various competing products are arranged on display in a room or hall and customers are asked to inspect the products and give their thoughts and reactions.

Usually, of course, customers will talk about products in terms of both general attributes and specific characteristics— observations ranging from 'It's easy to use' to 'I don't like the colour'. As in the performance specification method, it may be necessary to interpret the more general statements into more precise statements of requirements, but it is important to try to identify and to preserve the customers' wishes and preferences, rather than to reinterpret their observations into the designer's perceptions of what the customers 'really mean'. For this reason, words and phrases actually used by customers are often retained in statements of product attributes, even though they may seem to be vague and imprecise.

Determine the relative importance of the attributes

Of course, not all the identified product attributes will be equally important to customers. For example, 'easy to use' may be regarded as much more important than 'easy to maintain'. Also, some requirements (as noted in the performance specification method) may be 'demands' or absolute requirements (e.g. 'safe to use') rather than relative preferences.

The design team will want to know which attributes of their product design are ones that most heavily affect customers' perceptions of their product, and so it is necessary to establish the relative importance of those attributes to the customers themselves. Again market research methods can help to establish these relative preferences, and provide confirmation of whether what customers say they want is actually reflected in what they buy.

Some relatively simple techniques can also be used in order to assess the relative importance of the identified attributes. For example, customers can be asked to rank-order their statements of requirements, or to allocate 'points' (preferably from a fixed maximum points allowance) to the various attributes. (Techniques for doing this are discussed in the weighted objectives method, Chapter 9.)

The outcome of this step in the procedure is the allocation of relative 'weights' to the set of customer-specified product attributes. Normally, a percentage value is set for each attribute, i.e. the weights for the complete set of attributes add up to a total of 100.

Evaluate the attributes
of competing products

Customers often make judgements about product attributes in terms of comparisons with other products. For example, a car buyer may say that Car A 'feels more responsive than Car B'. This use of comparisons is perfectly understandable, given that customers are not usually experts and can only guess at what is possible in product design through observation of what some products actually achieve. Market research information is also often collected by methods of comparison between products.

In a competitive market, therefore, the design team has to try to ensure that its product will satisfy customer requirements better than the competitor products. The performance of the competition is therefore analyzed, particularly with regard to those product attributes that are weighted high in relative importance. Some of these performance measures will be objective and quantitative, whilst some will be subjective comparisons as made by customers. However, even when objective measures can be made, these should be checked against the customers' perceptions, which may not correspond with the objective measures.

In designing a new product, there may not be many competitor products, but that would be unusual; most product designs have to compete against existing products already on the market. In those cases where a design team is redesigning or improving an existing product, this step in the procedure not only highlights where improvements to the design team's product may be necessary, but also where this current product already has advantages over the competition, which should be maintained. The performance scores for the team's own current product and for the competition should be listed against the set of product attributes.

Draw a matrix of
product attributes
against engineering
characteristics

As suggested above, customers are not experts and therefore cannot usually specify their requirements in terms of the product's engineering characteristics that influence those requirements. For example, the car buyer may know what 'responsiveness' feels like, but is unlikely to be able to refer to this in terms of engine torque. It is therefore necessary for the design team to identify those engineering characteristics of their product that satisfy or influence in any way the customer requirements. For instance, the overall weight of a car, as well as its engine torque, will influence its 'responsiveness'.

The engineering characteristics must be real, measurable characteristics over which the engineering designer has some control. It is understandable for customers to be rather vague about their requirements, or to express them in frustratingly subjective terms, but the engineering designer can only work with the quantitative parameters of identifiable engineering characteristics. It is through the adjustment of the parameters of those characteristics that the designer influences the performance and/or the customer's perception of the product. Therefore it is often necessary to put considerable effort into identifying the relevant engineering characteristics and ensuring that each of these can be expressed in measurable units.

Of course, not all engineering characteristics affect all product attributes, and drawing up a matrix will enable the team to identify which characteristics do affect which attributes. It is usual to list the attributes, together with their relative weights, vertically, down the left edge of the matrix, and the characteristics horizontally, along the top edge. The attributes thus form the rows of the matrix, and the characteristics form the columns. Each cell of the matrix represents a potential interaction or relationship between an engineering characteristic and a customer requirement.

Down the right edge of the matrix can be listed the results of the evaluation of competing products, showing the scores achieved against the product attributes for the competing products and the design team's own current product. Along the bottom edge of the matrix is the usual place for recording the units of measurement of the engineering characteristics. If a product already exists and is being redesigned then the product's own values for these characteristics can also be inserted here, together with values achieved by competitor products.

Identify the relationships between engineering characteristics and product attributes

By checking through the cells of the matrix it is possible to identify where any engineering characteristic influences any product attribute. These relationships between characteristics and attributes will not all be of equal value. That is to say, some characteristics will have a strong influence on some attributes, whilst other characteristics might only have a weak influence.

The design team therefore works methodically through the matrix, and records in the matrix cells wherever a relationship occurs, and the strength of that relationship. Sometimes numbers are used to represent the strength of the relationship (e.g.

6 for a strong relationship, 3 for a medium-strength relationship, 1 for a weak relationship), or symbols can be used. When numbers are used, it is possible to enter a second value in each cell, which is the relative weight of the attribute multiplied by the strength of the relationship. The larger scores amongst these values enable the design team easily to identify where the adjustment of engineering characteristics will have a large influence on customers' overall perception of the product. However, unless accurate measures of the strength of the relationships can be established, it must be remembered that there is a spurious accuracy implied by the numbers.

Identify any relevant interactions between engineering characteristics

It is often the case that engineering characteristics interact with each other, particularly in terms of their influence on customers' perceptions of the product. For example, a more powerful engine is also likely to be heavier, thus increasing the vehicle weight, and so not necessarily increasing its perceived 'responsiveness'. These interactions can be either negative or positive.

A simple way of checking these interactions is to add another section to the interaction matrix. This new section is usually added on top of the existing matrix, and because it provides a triangular-shaped 'roof' to the matrix the resulting diagram is often referred to as the 'house of quality' (see the Worked Example, Figure 48).

Working through the 'roof' matrix enables a systematic check to be made of the interactions between the engineering characteristics, and whether these interactions are negative or positive. However, many assumptions may have to be made about the final design when completing the 'roof' matrix, and it should be remembered that changes in the design concept may result in changes in these interactions.

Set target figures to be achieved for the engineering characteristics

In following this method so far, the design team will already have gained substantial insight into their product design, including customer perceptions of their product and of competing products, and how the engineering characteristics of the product relate to customer requirements. In this step of the procedure, the team determines the targets that can be set for the measurable parameters of the engineering characteristics in order to satisfy customer requirements or to improve the product over its competitors.

Of course, in a competitive situation it is important to know what the competitors achieve on the characteristics of their product, so detailed investigation of competitor products may be necessary. The design team can then set targets for themselves which would be better than the competition. Sometimes it may be necessary to conduct trials with customers in order to determine what would be acceptable target figures to set. This is similar to determining values in a performance specification.

Summary

Aim

To set targets to be achieved for the engineering characteristics of a product, such that they satisfy customer requirements.

Procedure

1. Identify customer requirements in terms of product attributes.
It is important that 'the voice of the customer' is recognized, and that customer requirements are not subject to 'reinterpretation' by the design team.

2. Determine the relative importance of the attributes.
Techniques of rank-ordering or points-allocation can be used to help determine the relative weights that should be attached to the various attributes. Percentage weights are normally used.

3. Evaluate the attributes of competing products.
Performance scores for competing products and the design team's own product (if a version of it already exists) should be listed against the set of customer requirements.

4. Draw a matrix of product attributes against engineering characteristics.
Include all the engineering characteristics that influence any of the product attributes and ensure that they are expressed in measurable units.

5. Identify the relationships between engineering characteristics and product attributes.
The strength of the relationships can be indicated by either symbols or numbers; using numbers has some advantages, but can introduce a spurious 'accuracy'.

6. Identify any relevant interactions between engineering characteristics.

The 'roof' matrix of the 'house of quality' provides this check, but may be dependent upon changes in the design concept.

7. Set target figures to be achieved for the engineering characteristics.
Use information from competitor products or from trials with customers.

Examples

Example 1: Bicycle splashguard

This example is a relatively simple product, but it illustrates how considerable effort can be necessary in designing to satisfy customer requirements even for simple products. It is a design for a new product type—a detachable splashguard for the rear wheel of mountain bikes. Normally such bicycles have no mudguards, but for circumstances in which the rider does not wish to dirty his or her clothes (from water and mud thrown up at the rear), a detachable splashguard was thought to be a potentially desirable product.

Figure 44 shows the QFD interaction matrix prepared for this design problem. The design team interviewed mountain bicycle riders and asked them what features they would like to see incorporated in a detachable splashguard, and then organized this information (retaining 'the voice of the customer' as far as possible) into product attributes. Additional requirements from the sponsor of the project were also added, such as cost, time and manufacturing requirements. These are all listed, in groups, down the left edge of the matrix.

Some of the requirements were 'musts', i.e. they had to be satisfied absolutely, whereas others could be weighted relative to each other. These relative weights are listed alongside the requirements, and the absolute requirements are marked with an asterisk.

Since there were no comparable products already on the market, the team could not evaluate the attributes of competing products. However, they decided to make comparisons with two other possible means of avoiding dirt being splashed onto the rider—a fixed mudguard and a raincoat. The evaluations of these alternatives are given down the right edge of the matrix.

Engineering characteristics for a splashguard design were then established, related to the desired product attributes. For example, the 'easy to attach' attribute could be measured by (1)

| | ENGINEER REQUIREMENTS | | | | | | | | | | | | Bench marks | |

Bicycle splashgard

Customer Requirement			Weighting (total 100)	Steps to attach	Time to attach	Steps to detach	Time to detach	# of parts	# nonstd tools needed	# of std tools needed	Weight of all parts	Force to cause release	Bikes fit (currently on market)	Stiffness (lateral)	% of water blocked	Nonremovable fenders	Raincoat
Functional performance		Keeps water off rider	•												9	4	5
	Attach/detach	Easy to attach	7	9	3			3	3	9						1	4
		Easy to detach	4	3	9			3	3							1	5
		Fast to attach	3			9	3	3	3	9						1	4
		Fast to detach	1			3	9	3	3							1	5
		Can attach when bike is dirty	3					3	3							1	3
		Can detach when bike is dirty	1					3	3						/	1	5
	Interface with bike	Not mar	10	1	1	1	1			1						3	5
		Not catch water, etc.	7					9								3	5
	Structural integrity	Not rattle	8	1		1		3								3	3
		Not wobble	7	1		1		3						9		2	2
		Not bend	4	1		1		3						3		3	N/A
		Long life	11	1		1		3								3	2
		Lightweight	7					3			9					5	3
		Not release accidentally	10									9				5	N/A
Spatial constraints	Fit	Most bikes	7										9			4	5
	Not interfere	With rider	•													5	1
		With drive train	•													5	3
		With lights & generator	•													4	4
		With brakes	•													4	3
		With pannier	•													5	3
		With kickstand	•													5	5
	Appearance	Streamlined	5													2	1
		Popular color	5													5	3
	Time	3 months' development	•													/	5
		Marketable in 12 months	•													/	5
	Cost	Minimum capital <$15,000	•													/	5
		Manufacturing <$3 each	•													5	5
	Manufacture	200,000/year for 5 years	•													/	
		Use existing facilities	•													/	
		Units		#	sec	#	sec	#	#	#	oz	lb	%	in/lb	%		
		Targets		1	2	2	3	2	0	0	8	10	95	.01	95		
		Nonremovable fender		4	300	4	200	3	0	2	8	300	100	.01	95		
		Raincoat		3	15	3	5	1	0	0	6	?	100	Low	100		

CUSTOMERS REQUIREMENTS

Figure 44
Interaction matrix of requirements for the bicycle splashguard

the number of steps needed to attach, (2) the time needed to attach, (3) the number of parts needed, and (4) the number of standard tools needed. These characteristics (called 'engineer requirements' in the example) are listed along the top of the matrix, and values for the strength of the relationship between characteristics and attributes are shown in the appropriate cells of the matrix.

Finally, along the bottom of the matrix are listed the units by which the engineering characteristics can be measured, and the targets set for the new product design in comparison with those achieved by the alternatives. The design team thus established a thorough understanding of this novel design problem and determined a set of measurable targets to achieve in their design. (Source: Ullman.)

Example 2: Cordless drill

This example refers to a project to design a hand-held cordless drill for the professional market. It is used to demonstrate some of the principles of QFD, especially the 'house of quality' interaction matrices, shown in Figure 45.

Figure 45
'House of quality' for the design of a cordless drill

A simple list of product attributes is shown (here CAs, or 'customer attributes'), together with corresponding engineering

CAs	Relative Importance	+ Max. Power	+ Torque	+ Work Output	– Mass	+ Tool Life
Can use tool continuously	25			9	3	
Tool is powerful	25	9	9			
Tool is multi-purpose	20	6	6			
Tool is easy to handle	20				9	
Tool is long-lasting	25					9
Imputed Importance		1.5	1.5	1.0	1.1	1.0
Units		W	Nm	kJ	kg	hrs
Competitor Values		85	1.5	25	1.59	1000
Targets						

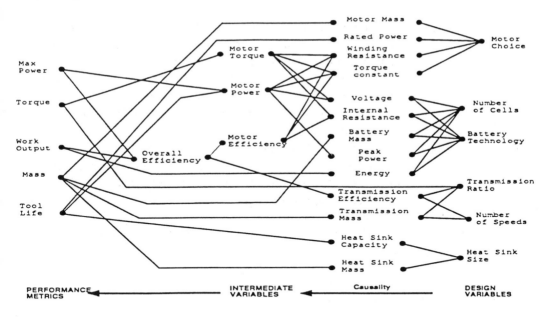

PERFORMANCE
METRICS

INTERMEDIATE
VARIABLES

Causality

DESIGN
VARIABLES

Figure 46
The network of physical
properties influencing the
engineering
characteristics of the
cordless drill

characteristics (ECs). Positive or negative signs are given with the ECs to indicate the preferred desirability of either increasing or decreasing the value of a characteristic. In the body of the matrix, values for the strengths of the relationships between attributes and characteristics have been inserted, and the 'roof' has been partially completed for the interactions between ECs. The supplementary information along the bottom of the matrix includes the EC values achieved by the principal competitor product, and the 'imputed importance' of each EC. This is a value derived by computing the weighted sums of EC–CA relationships. These relative values indicate where changing an engineering characteristic may have most influence.

However, the complexities of the engineering characteristics themselves are illustrated in Figure 46. This traces the network of properties which influence the engineering characteristics. For example, the 'torque' characteristic is determined by the transmission ratio and the motor torque, which in turn is determined by winding resistance, torque constant, voltage and internal resistance. Thus the 'performance metrics', or engineering characteristics, are determined by a complex network of design variables, or physical properties. It is in determining these basic variables that the engineering designer ultimately satisfies the customer's requirements. (Source: Ramaswamy and Ulrich.)

*Worked example: Car
door*

This example concentrates on selected attributes of just one component, a car door. Figure 47 shows the first stage, of developing and refining the set of product attributes from research on customer requirements. Using the objectives tree method enabled primary, secondary and tertiary levels of customer requirements to be identified and sorted into attribute 'bundles'. The relative importance weight of each attribute was also determined by market research surveys.

Using 'hall tests' customer perceptions of two competing products were established in comparison with their perceptions of the design team's own existing product. These customer perceptions were scored on a five-point scale, with a score of 5 representing the perceived best performance, and 1 representing the worst.

Figure 47
Objectives tree analysis
of customer requirements
for a car door

Part of the final, fully developed 'house of quality' is shown in Figure 48. The customer perceptions of the performance of competing products are shown graphically on the right.

Figure 48   'House of quality' (partial) for the car door

Objective measures of the relevant engineering characteristics were determined, and are shown below the matrix for the current and two competing products. Positive and negative interactions between ECs are shown in the matrix 'roof'. Finally, on the 'bottom line' are the targets set for a redesign of the car door, after considerations not only of imputed importance but also the technical difficulty and estimated cost of making improvements on the current design. (Source: Hauser and Clausing.)

8 *Generating Alternatives*

The generation of solutions is, of course, the essential, central aspect of designing. Whether one sees it as a mysterious act of creativity or as a logical process of problem solving, the whole purpose of design is to make a proposal for something new—something which does not yet exist.

The focus of much writing and teaching in design is therefore on novel products or machines, which often appear to have arisen spontaneously from the designer's mind. However, this overlooks the fact that most designing is actually a variation from or modification to an already-existing product or machine. Clients and customers usually want improvements rather than novelties.

Making variations on established themes is therefore an important feature of design activity. It is also the way in which much creative thinking actually develops. In particular, creativity can often be seen as the reordering or recombination of existing elements.

This creative reordering is possible because even a relatively small number of basic elements or components can usually be combined in a large number of different ways. A simple example of arranging adjacent squares into patterns demonstrates this:

No. of Squares	No. of distinct shape arrangements
2	1
3	2
4	5
5	12
6	35
7	108
8	369
.	.
.	.
.	.
16	13 079 255

The number of different arrangements, i.e. patterns or designs, soon becomes a 'combinatorial explosion' of possibilities.

The *morphological chart* method exploits this phenomenon, and encourages the designer to identify novel combinations of elements or components. The chart sets out the complete range of elements, components or sub-solutions that can be combined together to make a solution. The number of possible combinations is usually very high, and includes not only existing, conventional solutions but also a wide range of variations and completely novel solutions.

The main aim of this method is to widen the search for possible new solutions. 'Morphology' means the study of shape or form; so a 'morphological analysis' is a systematic attempt to analyze the form that a product or machine might take, and a 'morphological chart' is a summary of this analysis. Different combinations of sub-solutions can be selected from the chart, perhaps leading to new solutions that have not previously been identified.

The Morphological Chart Method

Procedure

List the features or functions that are essential to the product

The purpose of this list is to try to establish those essential aspects that must be incorporated in the product, or that it must be capable of doing. These are therefore usually expressed in rather abstract terms of product requirements or functions. In the morphological chart method they are sometimes called the design 'parameters'. As with many other design methods, instead of thinking in terms of the physical components that a typical product might have, you have to think of the functions that those components serve.

The items in the list should all be at the same level of generality, and they should be as independent of each other as possible. They must also comprehensively cover the necessary functions of the product or machine to be designed. However, the list must not be too long; if it is, then the eventual range of possible combinations of sub-solutions may become unmanageably large. About four to eight features or functions would make a sensible and manageable list.

For each feature or function list the means by which it might be achieved

These secondary lists are the individual sub-solutions which, when combined, one from each list, form the overall design solution. These sub-solutions can also be expressed in rather general terms, but it is probably better if they can be identified as actual components or physical embodiments. For instance, if one of the 'functions' of a vehicle is that it has motive power, then the different 'means' of achieving this might be engines using different fuels—e.g. petrol, diesel, electricity, gas.

The lists of means can include not only the existing, conventional components or sub-solutions of the particular product, but also new ones that you think might be feasible.

Draw up a chart containing all the possible sub-solutions

The morphological chart is constructed from the previous lists. At first, this is simply a grid of empty squares. Down the left-hand side are listed the essential features or functions of the product—i.e. the first list made earlier. Then across each row of the chart are entered the appropriate secondary lists of sub-solutions or means of achieving the functions. There is no relationship within the columns of the chart; the separate squares are simply convenient locations for the separate items. There might be, say, three means of achieving the first function, five means of achieving the second function, two means of achieving the third, and so on.

When it is finished, the morphological chart contains the complete range of all the theoretically-possible different solution forms for the product. This complete range of solutions consists of the combinations made up by selecting one sub-solution at a time from each row. The total number of combinations is therefore often very large. For instance, if there were only three rows (functions), with three squares (means) in the first row, five in the second, and two in the third, then the complete set of possible combinations would number $3 \times 5 \times 2 = 30$. Because of this potential combinatorial explosion, the list of means for each function should be kept reasonably short.

Identify feasible combinations of sub-solutions

Clearly, for any product the complete range of possible combinations can be a very large number. Some of these combinations—probably a small number—will be existing solutions; some will be feasible new solutions; and some—possibly a great number—will be impossible solutions, for reasons of practicality or because particular pairs of sub-solutions may be incompatible.

If the total number of possible combinations is not too large, then it may be possible to list each combination, and so to set out the complete range of solutions. Each potential solution can then be considered, and one or more of the better solutions (for reasons of cost, performance, novelty, or whatever criteria are important) chosen for further development.

If—as is more likely—the total number of possible combinations is very large, then some means has to be found of reducing this to something more manageable. One way of doing this is to choose only a restricted set of sub-solutions from each row— say, those that are known to be efficient or practical, or look promising for some other reason. Another way is to identify the non-feasible sub-solutions, or incompatible pairs of sub-solutions, and so rule out those combinations that would include them.

A really exhaustive search of all the possible combinations in a morphological chart requires much patient and tedious work. (It is perhaps something where computer aids might help.) The only alternative is a more intuitive—or perhaps random— search of the chart for solutions.

Summary

Aim

To generate the complete range of alternative design solutions for a product, and hence to widen the search for potential new solutions.

Procedure

1. List the features or functions that are essential to the product. Whilst not being too long, the list must comprehensively cover the functions, at an appropriate level of generalization.

2. For each feature or function list the means by which it might be achieved.
These lists might include new ideas as well as known existing components or sub-solutions.

3. Draw up a chart containing all the possible sub-solutions. This morphological chart represents the total solution space for the product, made up of the combinations of sub-solutions.

4. Identify feasible combinations of sub-solutions.

The total number of possible combinations may be very large, and so search strategies may have to be guided by constraints or criteria.

Examples

Example 1: Vehicle configurations

A simple example of morphological analysis applied to vehicle configurations arose in the course of the design of vehicles for land and water speed record attempts, where it was necessary to define acceptable vehicle configurations. The functions of such vehicles were defined as:

Support
Drive
Stability
Control

In each case, the means of achieving these functions were classified as land, water or air. The resulting morphological chart is shown in Figure 49 (overleaf). You might like to try identifying other combinations than those defining a car, a boat or an aeroplane, e.g. a land yacht, a windsurf board, a hover-craft or a hydrofoil. (Source: Norris.)

Example 2: Potato harvesting machine

An improvement to the often rather abstract and wordy form of morphological charts can be made by using—where possible—pictorial representations of the different means for achieving the functions. An example is shown in Figure 50—a morphological chart for a potato harvesting machine. One selected combina-tion of sub-solutions is highlighted in the chart. Notice that two sub-solutions are chosen for the sub-function 'separate stones'. This suggests either that each of these two sub-solutions is not really adequate on its own, or else that the morphological chart itself has not really been constructed carefully enough—perhaps 'separate stones' is not just one sub-function, but needs to be more carefully defined. (Source: Pahl and Beitz.)

Example 3: Welding positioner

A welding positioner is a device for supporting and holding a workpiece and locating it in a suitable position for welding. Figure 51 shows a morphological chart for such a device, using

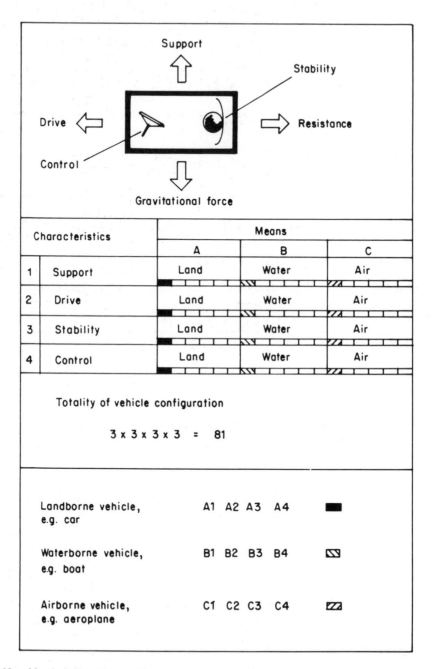

Figure 49   Morphological chart of basic vehicle configurations

Figure 50
Morphological chart for a potato harvesting machine. The shaded boxes show one possible combination of sub-solutions

Sub-functions	Solutions 1	2	3	4	...
1 Lift	and pressure roller	and pressure roller	and pressure roller	Pressure roller	...
2 Sift	Sifting belt	Sifting grid	Sifting drum	Sifting wheel	
3 Separate leaves			Plucker	...	...
4 Separate stones					...
5 Sort potatoes	by hand	by friction (inclined plane)	checksize (hole gauge)	check mass (weighing)	...
6 Collect	Tipping hopper	Conveyor	Sack-filling device		

Combination of principles

Figure 51
Morphological chart for a welding positioner, with one possible combination of sub-solutions picked out by the zig-zag line

Partial Functions	Action Principles/Families of Function-carriers 1	2	3	4	5	6
1 ENABLE connection with workpiece	Form interlocking	mechanical — screw or bolt	wedge	Force locking (friction) — pneumatic	hydraulic	magnetic
2 ENABLE rotational movement	Rotational guidance / sliding journal bearing	rolling bearing				
3 ENABLE tilting movement	cylinder	sphere	fulcrum pin position	hang from above		
4 ENABLE height adjustment	straight line guidance, bearing, sliding or rolling	screw thread		lever mechanism		
21 31 41 LOCK state	hold directly by hand, weight of workpiece	form interlocking / hole - pin	ratchet mechanism	force locking (friction) / within guidance	screw screw with washer	wedge, brake block
22 32 42 DRIVE (by hand)	Direct	with mechanical advantage device / gear wheel pair	rack and pinion	helical gears (crossed)	worm and worm-wheel	band, rope, chain / lever, eccentric cam
23 33 43 CONTROL of movement	through drive mechanism	through locking (ratchet)				
Show position	mechanical / line scale	pointer scale	optical	electronic	mechanical stop	

Concept		B	C	D
Note: Function-carriers for functions 2.1., 3.1., 4.1., 2.2., 3.2., 4.2., 2.3., 3.3., 4.3., are compatible with all combinations of the main functions	*(concept sketches)*			
Combinations of Function-carriers — Function 2	1	1	1	1
Function 3	1	3 (2 pins)	3 (lever)	3 (1 pin)
Function 4	1	1	1	1
Rough evaluation — Technical (weak points)		about equal value (Function 3.1)	(Function 3.1., 3.2)	(Function 3.1)
manufacturing costs	2	3	3	4
Overall rank order	4	2	3	1

Figure 52
Four possible combinations for the welding positioner worked up into concept sketches

words augmented by sketch diagrams. One possible combination of sub-solutions is indicated by the zig-zag line through the chart. Even then, it was found that there were alternatives for the actual embodiment of some of the sub-solutions. For example, the sketches in Figure 52 show alternative configurations for the chosen means of enabling the tilting movement. (Source: Hubka.)

Example 4: Shaft coupling

This example shows that even small components can be usefully subject to morphological analysis. The example is that of a shaft coupling similar to the conventional 'Oldham' coupling, which transmits torque even in the case of radial and axial offsets of the shafts. Figure 53 shows a part of the morphological chart that was drawn up. One solution type (A) was analyzed into its components and elements (presented here in columns, rather than in rows) and the various sub-solutions listed in pictures and words. Two new alternative combinations (B and C) are shown by the different sets of dots in the squares of the chart. One of these (B) was developed and patented as a novel design, as shown in Figure 54. (Source: Ehrlenspiel and John.)

Figure 53
Part of a morphological
chart for a shaft coupling

Figure 54
Design for a novel form of
coupling, derived from
one of the combinations
in the morphological chart

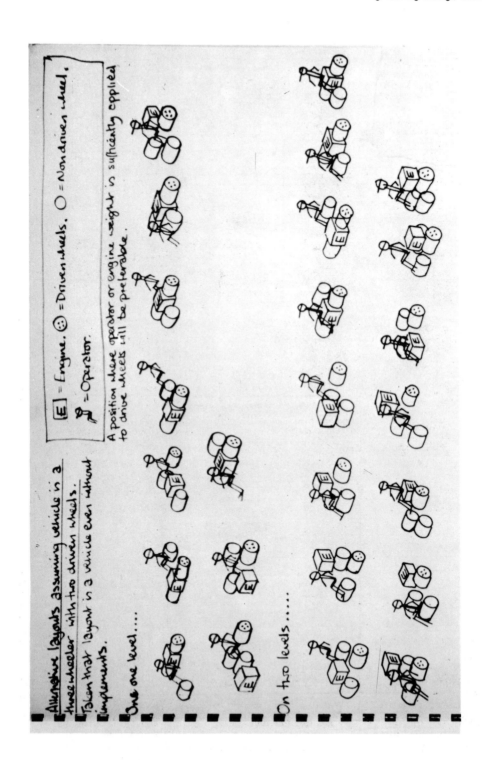

Alternative layouts assuming vehicle is a three-wheeler with two driven wheels.

Taken that layout in a vehicle even without implements.

E = Engine. ⊙ = Driven wheels. ○ = Non driven wheel.
☆ = Operator.

A position where operator or engine weight is sufficiently applied to drive wheels will be preferable.

On one level......

On two levels......

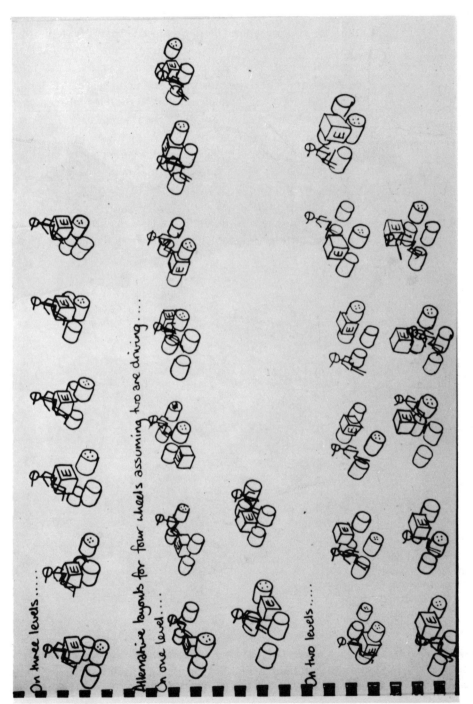

Figure 55   Morphological analysis of a field maintenance machine

Example 5: Field
maintenance machine

This example shows an adaptation of the principles of morphological analysis. It is concerned solely with the formal arrangement, or configuration, of the essential basic elements of the product, and represents the alternative configurations in purely graphical terms. The example is that of a sports field maintenance machine, and the morphological analysis shown in Figure 55 (see previous pages) explores the alternatives for configuring the elements of operator, engine, driven wheels and non-driven wheels, and the possible disposition of these on one, two or three vertical levels. The sketches show alternative layouts for either a three-or four-wheel machine, with the arrangement options systematically varied so as to generate all possible design configurations.

The options were evaluated against design criteria, and one preferred option developed, as shown in Figure 56. (Source: Tovey.)

Figure 56
Design sketch for one
preferred option derived
from the morphological
analysis

Worked example:
Forklift truck

This example is concerned with finding alternative versions of the conventional forklift truck used for lifting and carrying loads in factories, warehouses, etc. If we investigate a few of these machines we might identify the essential generic features as follows:

1. Means of support which allows movement
2. Means of moving the vehicle
3. Means of steering the vehicle
4. Means of stopping the vehicle
5. Means of lifting loads
6. Location for operator

These features seem to be common to all forklift trucks, although different versions have different means of achieving the functions. For example, most such trucks run on wheels (means of support) that allow the vehicle to go anywhere on a flat surface, but some are constrained to run on rails.

When we look at the means of moving the vehicle, we might conclude that this is too general a feature and we decide that it should be broken-down into separate features for (a) the means of propulsion (normally driven wheels), (b) the power source (such as electric motor, petrol or diesel engine) and (c) transmission type (gears and shafts, belt, hydraulic, etc.). Adding some new and perhaps rather fanciful alternatives to the conventional alternatives would enable a list like this to be generated:

Feature	Means
Support	Wheels, track, air cushion, slides, pedipulators
Propulsion	Driven wheels, air thrust, moving cable, linear induction
Power	Electric, petrol, diesel, bottled gas, steam
Transmission	Gears and shafts, belts, chains, hydraulic, flexible cable
Steering	Turning wheels, air thrust, rails
Stopping	Brakes, reverse thrust, ratchet
Lifting	Hydraulic ram, rack and pinion, screw, chain or rope hoist
Operator	Seated at front, seated at rear, standing, walking, remote control

A morphological chart incorporating these lists is shown in Figure 57. You might like to calculate how many possible different solution combinations there are in this chart.

There are a staggering 90 000 possible forklift truck designs in the chart. Of course, some of these are not practicable solutions, or imply incompatible options—for example, an air cushion vehicle could not have steering by wheels. A typical, conventio-

Feature	Means				
Support	Wheels	Track	Air cushion	Slides	Pedipulators
Propulsion	Driven wheels	Air thrust	Moving cable	Linear induction	
Power	Electric	Petrol	Diesel	Bottled gas	Steam
Transmission	Gears and shafts	Belts	Chains	Hydraulic	Flexible cable
Steering	Turning wheels	Air thrust	Rails		
Stopping	Brakes	Reverse thrust	Ratchet		
Lifting	Hydraulic ram	Rack and pinion	Screw	Chain or rope hoist	
Operator	Seated at front	Seated at rear	Standing	Walking	Remote control

Figure 57
Morphological chart for
forklift trucks

nal forklift truck would comprise the following set of options from the chart:

Support	Wheels
Propulsion	Driven wheels
Power	Diesel engine
Transmission	Gears and shafts
Steering	Turning wheels
Stopping	Brakes
Lifting	Rack and pinion
Operator	Seated at rear

The inclusion of a few unconventional options in the chart suggests some possibilities for radical new designs. For instance, the idea of 'pedipulators' (i.e. walking mechanisms similar to legs and feet) might lead to designs suitable for use on rough ground such as building sites—or even capable of ascending flights of steps.

The chart can also be used to help generate somewhat less fanciful but nonetheless novel design ideas. For example, the idea of using rails for steering might well be appropriate in some large warehouses, where the rails could be laid in the aisles between storage racks. The vehicle would have wheels for

Feature	Means				
Support	(Wheels)	Track	Air cushion	Slides	Pedipulators
Propulsion	(Driven wheels)	Air thrust	Moving cable	Linear induction	
Power	(Electric)	Petrol	Diesel	Bottled gas	Steam
Transmission	Gears and shafts	(Belts)	Chains	Hydraulic	Flexible cable
Steering	Turning wheels	Air thrust	(Rails)		
Stopping	(Brakes)	Reverse thrust	Ratchet		
Lifting	Hydraulic ram	Rack and pinion	(Screw)	Chain or rope hoist	
Operator	Seated at front	Seated at rear	Standing	Walking	(Remote control)

Figure 58
One selected combination of sub-solutions from the morphological chart

support and for providing propulsion. It would be electrically powered since it would be used indoors. One of the problems of electric vehicles is the limited battery power, so we might propose that our new design would pick up power from a live electric rail—like subway trains. This might be feasible in a fully-automated warehouse which would not have the safety problems associated with people having to cross the rails. The 'operator' feature would therefore be remote control. A compatible set of sub-solutions for this new design therefore becomes:

Support	Wheels
Propulsion	Driven wheels
Power	Electric motor
Transmission	Belt
Steering	Rails
Stopping	Brakes
Lifting	Screw
Operator	Remote control

This set is shown as a selection from the morphological chart in Figure 58.

9 *Evaluating Alternatives*

When a range of alternative designs has been created, the designer is then faced with the problem of selecting the best one. At various points in the design process there may also be decisions of choice to be made between alternative sub-solutions or alternative features that might be incorporated into a final design. Choosing between alternatives is therefore a common feature of design activity.

Choices can be made by guesswork, by 'intuition', by experience, or by arbitrary decision. However, it is better if a choice can be made by some more rational, or at least open, procedure. Not only will the designer feel more secure in making the choice, but others involved in decision making, such as clients, managers and colleagues in the design team, will be able to participate in, or assess the validity of the choice.

If some of the previous design methods have already been used in the design process, then there should be some information available which should guide a choice between alternatives. For example, design proposals can be checked against criteria established by the performance specification method; and if design objectives have been established by the objectives tree method then these can be used in the evaluation of alternative designs.

In fact, the evaluation of alternatives can only be done by considering the objectives that the design is supposed to achieve. An evaluation assesses the overall 'value' or 'utility' of a particular design proposal with respect to the design objectives. However, different objectives may be regarded as having different 'values' in comparison with each other, i.e. may be regarded as being more important. Therefore it usually becomes necessary to have some means of differentially 'weighting' objectives, so that the performances of alternative designs can be assessed and compared across the whole set of objectives.

The *weighted objectives* method provides a means of assessing and comparing alternative designs, using differentially weighted objectives. This method assigns numerical weights to objectives, and numerical scores to the performances of alternative designs measured against these objectives. However, it must be emphasized that such weighting and scoring can lead the unwary into some very dubious arithmetic. Simply assigning numbers to objectives, or objects, does not mean that arithmetical operations can be applied to them. For instance, a football player assigned the number 9 is not necessarily three times as good, or worth three times as much as a player assigned the number 3—even though he may score three times as many goals! Arithmetical operations can only be applied to data which have been measured on an *interval* or *ratio* scale.

The Weighted Objectives Method

Procedure

List the design objectives

In order to make any kind of evaluation it is necessary to have a set of criteria, and these must be based on the design objectives, i.e. what it is that the design is meant to achieve. These objectives should have been established at an early point in the design process. However, at the later stages of the process—when evaluation becomes especially important—the early set of objectives may well have become modified, or may not be entirely appropriate to the designs that have actually been developed. Some clarification of the set of objectives may therefore be necessary as a preliminary stage in the evaluation procedure.

The objectives will include technical and economic factors, user requirements, safety requirements, and so on. A comprehensive list should be drawn up. Wherever possible, an objective should be stated in such a way that a quantitative assessment can be made of the performance achieved by a design on that objective. Some objectives will inevitably relate to qualitative aspects of the design; these may later be allocated 'scores', but the earlier warning about limitations on the use of arithmetic must be remembered.

The list of objectives will contain a wide variety of design requirements, some of which will be considered to be more important than others. As a first step towards determining relative 'weights' for the objectives, it is usually possible to list them in a rank order of importance. One way of doing this is to write each objective on a separate card and then to sort the cards into a comparative rank order, i.e. from 'most important' to 'least important'.

As with many other aspects of this design method, it is usually helpful if the rank-ordering of objectives can be done as a team effort, since different members of a design team may well give different priorities to different objectives. Discussion of these differences will (hopefully!) lead to a team consensus. Alternatively, the client may be asked to decide the rank-ordering, or market research might be able to provide customers' preferences.

The rank-ordering process can be helped by systematically comparing pairs of objectives, one against the other. A simple chart can be used to record the comparisons and to arrive at a rank order, like this:

Objectives	A	B	C	D	E	Row totals
A	—	0	0	0	1	1
B	1	—	1	1	1	4
C	1	0	—	1	1	3
D	1	0	0	—	1	2
E	0	0	0	0	—	0

Each objective is considered in turn against each of the others. A figure 1 or 0 is entered into the relevant matrix cell in the chart, depending on whether the first objective is considered more or less important than the second, and so on. For example, start with objective A and work along the chart row, asking 'Is A more important than B?'... 'than C?'... 'than D?', etc. If it is considered more important, a 1 is entered in the matrix cell; if it is considered less important, a 0 is entered. In the example above, objective A is considered less important than all others except objective E.

As each row is completed, so the corresponding column can also be completed with an opposite set of figures; thus, if row A reads 0001 then column A must be 1110. If any pair of objectives is considered equally important, a 1/2 can be entered in both relevant squares.

When all pairs of comparisons have been made, the row totals indicate the rank order of objectives. The highest row total indicates the highest priority objective. In the example above, the rank order therefore emerges as:

B
C
D
A
E

It is here that one of the first problems of ranking may emerge, where relationships may not turn out to be transitive. That is, objective A may be considered more important than objective B, and objective B more important than objective C, but objective C may then be considered more important than objective A. Some hard decisions may have to be made to resolve such problems!

A rank-ordering is an example of an *ordinal* scale; arithmetical operations cannot be performed on an ordinal scale.

Assign relative weightings to the objectives

The next step is to assign a numerical value to each objective, representing its 'weight' relative to the other objectives. A simple way of doing this is to consider the rank-ordered list as though the objectives are placed in positions of relative importance, or value, on a scale of, say, 1 to 10 or 1 to 100. In the example above, the rank-ordered objectives might be placed in relative positions on a scale of 1–10 like this:

10 B
9
8
7 C
6
5 D
4 A
3
2 E
1

The most important objective, B, has been given the value 10, and the others then given values relative to this. Thus, objective

C is valued as about 70% of the value of objective B; objective A is valued twice as highly as objective E; etc. The corresponding scale values are the relative weights of the objectives. (Note that the highest and lowest ranked objectives are not necessarily placed at the absolute top and bottom positions of the scale.)

If you can achieve such relative weightings, and feel confident about the relative positions of the objectives on the scale, then you have converted the ordinal rank-order scale into an *interval value scale*, which can be used for arithmetic operations.

An alternative procedure is to decide to share a certain number of 'points'—say, 100—amongst all the objectives, awarding points on relative value and making trade-offs and adjustments between the points awarded to different objectives until acceptable relative allocations are achieved. This can be done on a team basis, with members of the team each asked to allocate, or 'spend', a fixed number of total points between the objectives according to how highly they value them. If 100 points were allocated amongst objectives A to E in the earlier example, the results might be:

B   35
C   25
D   18
A   15
E   7

An objectives tree can be used to provide what is probably a more reliable method of assigning weights. The highest-level, overall objective is given the value 1.0; at each lower level, the sub-objectives are then given weights relative to each other but which also total 1.0. However, their 'true' weights are calculated as a fraction of the 'true' weight of the objective above them.

This is clarified by Figure 59 (overleaf). Each box in the tree is labelled with the objective's number (O_1, O_{11}, etc.), and given two values: its value relative to its neighbours at the same level, and its 'true' value or value relative to the overall objective. Thus, in Figure 59, objectives O_{111} and O_{112} (sub-objectives of O_{11}) are given values relative to each other of 0.67:0.33; but their 'true' values can only total 0.5 (the 'true' value of objective O_{11}) and are therefore calculated as $0.67 \times 0.5 = 0.34$ and $0.33 \times 0.5 = 0.16$.

Using this procedure it is easier to assign weights with some consistency because it is relatively easy to compare sub-

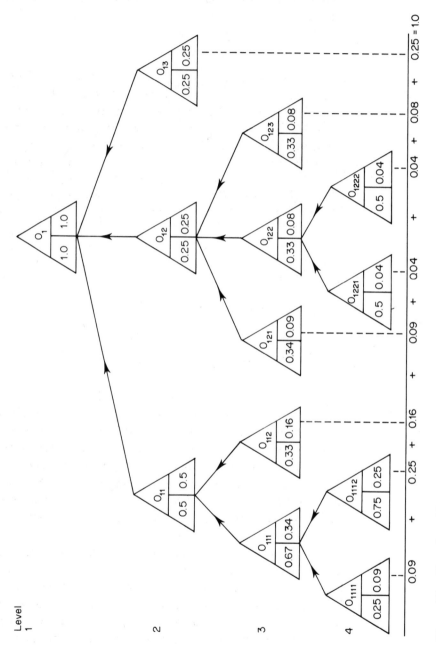

Figure 59 Use of an objectives tree for assigning relative weights to sub-objectives

objectives in small groups of two or three and with respect to a single higher-level objective. All the 'true' weights add up to 1.0, and this also ensures the arithmetical validity of the weights.

Establish performance parameters or utility scores for each of the objectives

It is necessary to convert the statements of objectives into parameters that can be measured, or at least estimated with some confidence. Thus, for instance, an objective for a machine to have 'high reliability' might be converted into a performance parameter of 'breakdowns per 10 000 hours running time', which might be either measured from available data or estimated from previous experience with that type of machine.

Some parameters will not be measurable in simple, quantifiable ways, but it may be possible to assign utility scores estimated on a points scale. The simplest scale usually has five grades, representing performance that is

Far below average
Below average
Average
Above average
Far above average

Often, a five-point scale (0–4) is too crude, and you will need to use perhaps a nine-point (0–8) or eleven-point (0–10) scale. The degrees of performance assessed by an eleven-point and a five-point scale might be compared as in Table 1 (overleaf).

Both quantitative and qualitative parameters can be compared together on a points scale, representing the worst to best possible performance range. For example, the fuel consumption and, say, the comfort of a motorcar could be represented on a seven-point scale as in Table 2.

Care must be taken in compiling such points scales, because the values ascribed to the parameters may not rise and fall linearly. For example, on the scale above, the value of decreasing fuel consumption is assumed to increase linearly, but it might well be regarded as more valuable to provide improvements in fuel consumption at the lower end of the scale rather than the upper end. That is, the 'utility curve' for a parameter might be an exponential or other curve, rather than linear.

Table 1

Eleven-point scale	Meaning	Five-point scale	Meaning
0	totally useless solution	0	inadequate
1	inadequate solution		
2	very poor solution	1	weak
3	poor solution		
4	tolerable solution	2	satisfactory
5	adequate solution		
6	satisfactory solution		
7	good solution	3	good
8	very good solution		
9	excellent solution	4	excellent
10	perfect or ideal solution		

Table 2

Points	Fuel consumption (miles/gal)	Comfort
0	<27	very uncomfortable
1	29	poor comfort
2	32	below average comfort
3	35	average comfort
4	38	above average comfort
5	41	good comfort
6	>43	extremely comfortable

*Calculate and compare
the relative utility
values of the
alternative designs*

The final step in the evaluation is to consider each alternative design proposal and to calculate for each one a score for its performance on the established parameters. Once again, the participation of all members of the design team is recommended (and especially those whose views ultimately count, such as customers!), since different solutions may be scored differently by different people.

The raw performance measures or points scores on each parameter for each alternative design must be adjusted to take account of the different weights of each objective. This is done by simply multiplying the score by the weight value, giving a set of adjusted scores for each alternative design which indicates the relative 'utility value' of that alternative for each objective.

These utility values are then used as a basis of comparison between the alternative designs. One of the simplest comparisons that can be made is to add up the utility value scores for each alternative. These total scores then allow the alternatives to be ranked in order of overall performance.

Other comparisons are possible, such as drawing graphs or histograms to represent the utility value 'profiles' of the alternative designs. These visual, rather than numerical, comparisons present a 'picture' which may be easier to absorb and reflect on. They also highlight where alternatives may be significantly different from each other in their performance.

The benefit of using this evaluation method often lies in making such comparisons between alternatives, rather than using it simply to try to choose the 'best' alternative. Many rather contentious weightings, points scores and other decisions will probably have been made in compiling the evaluation, and some of the arithmetic may well be highly dubious. The 'best' overall utility value may therefore be highly misleading; but the discussions, decisions, rankings and comparisons involved in the evaluation are certain to have been illuminating.

Summary

Aim

To compare the utility values of alternative design proposals, on the basis of performance against differentially weighted objectives.

Procedure

1. List the design objectives.
These may need modification from an initial list; an objectives tree can also be a useful feature of this method.

2. Rank order the list of objectives.
Pair-wise comparisons may help to establish the rank order.

3. Assign relative weightings to the objectives.
These numerical values should be on an interval scale; an alternative is to assign relative weights at different levels of an objectives tree, so that all weights sum to 1.0.

4. Establish performance parameters or utility scores for each of the objectives.
Both quantitative and qualitative objectives should be reduced to performance on simple points scales.

5. Calculate and compare the relative utility values of the alternative designs.
Multiply each parameter score by its weighted value—the 'best' alternative has the highest sum value; comparison and discussion of utility value profiles may be a better design aid than simply choosing the 'best'.

Examples

Example 1: Motorcar

The basic principles of the weighted objectives method are demonstrated in this simple example of the calculation of utility values for three different motorcars. Given that all three are about the same purchase price, the potential purchaser might set the following objectives:

Low fuel consumption
Low cost of spare parts
Easy to maintain
High comfort

These are considered to have the respective relative weights of 0.5, 0.2, 0.1 and 0.2.

The objective of 'low fuel consumption' is easily assigned a performance parameter—that of the known figure for fuel consumption from standard tests, in miles per gallon. The figures are converted in the comparison table (Figure 60) to points on a seven-point scale, as suggested in the Procedure.

Objective	Weight	Parameter	Car A			Car B			Car C		
			Magnitude	Score	Value	Magnitude	Score	Value	Magnitude	Score	Value
Low fuel consumption	0.5	Miles per gallon	33	2	1.0	40	4	2.0	36	3	1.5
Low cost of spare parts	0.2	Cost of 5 typical parts	£18	7	1.4	£22	5	1.0	£28	2	0.4
Easy to maintain	0.1	Simplicity of servicing	Very simple	5	0.5	Com-plicated	2	0.2	Average	3	0.3
High comfort	0.2	Comfort rating	Poor	2	0.4	Very good	5	1.0	Good	4	0.8
Overall utility value					3.3			4.2			3.0

Figure 60 Weighted objectives evaluation chart for three alternative motorcars

To measure 'cost of spare parts' it was decided to take the cost of a standard set of typical parts that might have to be purchased in, say, the first three years of a car's life. Again, a seven-point scale was used to convert the cost figures to a utility points score.

The objective of 'easy to maintain' was assessed on a seven-point scale of 'simplicity of servicing'. The assessment was based on the number of routine service tasks and the ease with which these could be done for each particular motorcar.

Finally, comfort was also assessed on a seven-point scale, similar to that presented in the Procedure.

For each alternative, the utility score for each objective is multiplied by the objective's weight, giving a relative utility value. If these are added together, an overall utility value for each alternative is obtained. As the figures in the chart of Figure 60 show, Car B emerges as clearly the 'best' overall.

Example 2: Test rig

This example is taken from the project for the design of a laboratory rig for carrying out impulse-load tests on shaft connectors. A thorough evaluation was made of a number of alternative designs, based on the objectives tree which was presented earlier in the objectives tree method (Example 3, Figure 23).

The objectives and sub-objectives at different levels were weighted in the manner described in the Procedure (see Figure 61). The design team then went on to devise measurable or assessable parameters for all of the objectives, as indicated in the comparison chart (Figure 62).

Utility values were calculated for each objective, for each of four alternative designs. The second alternative (variant V2) emerges as the 'best' solution, with an overall utility value of 6.816. However, variant V3 seems quite comparable, with an overall utility value of 6.446. A comparison of the 'value profiles' of these two alternatives was therefore made. This is shown in Figure 63, where the height of each bar in the chart represents the relative weight of each objective, and its length represents the score for that objective achieved by the particular design.

The chart shows that V2 has a more consistent profile than V3, with fewer relatively weak spots in its profile. V2 therefore seems to be a good, 'all-round' design, and the comparison confirms it as being the best of the alternatives. However,

Figure 61
Weighted values assigned to an objectives tree impulse-load testing machine

No.	Evaluation criteria	Wt.	Parameters	Unit	Variant V_1 Magn. m_{i1}	Value V_{i1}	Weighted value WV_{i1}	Variant V_2 Magn. m_{i2}	Value V_{i2}	Weighted value WV_{i2}	Variant V_3 Magn. m_{i3}	Value V_{i3}	Weighted value WV_{i3}	Variant V_4 Magn. m_{i4}	Value V_{i4}	Weighted value WV_{i4}
1	Low wear of moving parts	0.056	Amount of wear	-	high	3	0.168	low	6	0.336	average	4	0.224	low	6	0.336
2	Low susceptibility to vibrations	0.14	Natural frequency	s^{-1}	410	3	0.420	2370	7	0.980	2370	7	0.980	<410	2	0.280
3	Few disturbing factors	0.084	Disturbing factors	-	high	2	0.168	low	7	0.588	low	6	0.504	(average)	4	0.336
4	Tolerance of overloading	0.12	Overload reserve	%	5	5	0.600	10	7	0.840	10	7	0.840	20	8	0.960
5	High mechanical safety	0.21	Expected mechan. safety	-	average	4	0.840	high	7	1.470	high	7	1.470	very high	8	1.680
6	Few possible operator errors	0.09	Possibilities of operator errors	-	high	3	0.270	low	7	0.630	low	6	0.540	average	4	0.360
7	Small number of components	0.03	No. of components	-	average	5	0.150	average	4	0.120	average	4	0.120	low	6	0.180
8	Low complexity of components	0.012	Complexity of components	-	low	6	0.072	low	7	0.084	average	5	0.060	high	3	0.036
9	Many standard and bought-out parts	0.018	Proportion of standard and bought-out components	-	low	2	0.036	average	6	0.108	average	6	0.108	high	8	0.144
10	Simple assembly	0.04	Simplicity of assembly	-	low	3	0.120	average	5	0.200	average	5	0.200	high	7	0.280
11	Easy maintenance	0.06	Time and cost or maintenance	-	average	4	0.240	low	8	0.480	low	7	0.420	high	3	0.180
12	Quick exchange of test connections	0.084	Estimated time needed to exchange test connections	min	180	4	0.336	120	7	0.588	120	7	0.588	180	4	0.336
13	Good accessibility of measuring systems	0.056	Accessibility of measuring systems	-	good	7	0.392	good	7	0.392	good	7	0.392	average	5	0.280
		Σw_i = 1.0				$OV_1 = 51$ $R_1 = 0.39$	$OWV_1 =$ 3.812 $WR_1 = 0.38$		$OV_2 = 85$ $R_2 = 0.65$	$OWV_2 =$ 6.816 $WR_2 = 0.68$		$OV_3 = 78$ $R_3 = 0.60$	$OWV_3 =$ 6.446 $WR_3 = 0.64$		$OV_4 = 68$ $R_4 = 0.52$	$OWV_4 =$ 5.388 $WR_4 = 0.54$

Figure 62 Completed evaluation chart for four alternative designs for the impulse-load testing machine

Variant V_2: OWV_2 = 6.82   Variant V_3: OWV_3 = 6.45

Figure 63
Value profiles for
alternative test rig
designs V_2 and V_3

improvement of V3 in perhaps just one or two of its lower-scoring parameters might easily push it into the lead. (Source: Pahl and Beitz.)

Example 3: Swivel joint

This problem was the design of a swivel joint used in an underwater marine environment as part of a current-metering system. A previous design was considered unsuitable because of its high cost and poor performance (especially the high friction between adjacent moving parts).

Three different new designs were developed (Figure 64, overleaf), and evaluated by weighted objectives. Figure 65 shows the evaluation chart; each design is scored on a 0–10 scale for each objective, and each score multiplied by a weighting factor to give a utility value. Because of the nature of the problem, cost factors were given relatively high weights, whilst strength was low-weighted. This was because the operational loads were not severe, and the materials were chosen more for resistance to corrosion than for loading stresses.

Designs 1 and 3 achieved similar overall utility values, but design 1 was considered less complex in manufacture and was therefore chosen in preference to design 3. (Source: Shahin.)

Example 4: City car

This example continues the problem of designing a small car for use in the city or over a very limited journey range, introduced in the performance specification method (Example 4). As a part of the design study, the designers drew up a morphological chart of six basic types of city car and the variants within each type for aspects such as the positioning of the engine. This total

Figure 64   Three alternative designs for the swivel joint

Design criteria	Weight	Design 1		Design 2		Design 3	
	W^*	S	U	S	U	S	U
1 Cost							
Materials	6	8.5	0.51	5.5	0.33	7	0.42
Seals	2	8	0.16	8	0.16	8	0.16
Bearings	4	9	0.36	5	0.2	8	0.28
Washers	1	7.5	0.07	7.5	0.07	7.5	0.07
Squeeze packing	2	9	0.18	9	0.18	9	0.18
Bolts	1	9	0.09	9	0.09	8	0.08
Labour	6	8	0.48	5	0.3	7.5	0.45
Tools and equipment	6	8	0.48	5	0.3	7.5	0.45
Indirect cost	20	8.5	1.7	7	1.4	7.5	1.5
Marketing	2	7	0.14	8	0.16	9	0.18
2 Performance							
Sealing	9	8	0.72	8	0.72	8	0.72
Smoothness	9	5	0.45	9	0.81	8.5	0.76
Alignment	6	5	0.3	7	0.42	8	0.48
Growth formation	2	8	0.16	8	0.16	8	0.16
Maintenance	4	8	0.32	8	0.32	8	0.32
3 Manufacturing							
Ease	5	8.5	0.42	7	0.35	7.5	0.37
Time	5	9	0.45	4.5	0.22	7.5	0.37
Assembly	5	9	0.45	6.5	0.32	8	0.4
4 Strength	5	8	0.4	9.5	0.47	9.5	0.47
The overall utility			7.84		6.98		7.82

$^*W$ = percentage weight of each criterion (from 100)
S = score of quality of each design (from 10)
U = utility (weighted score) of design = $W \times S$

Figure 65
Evaluation chart for the
three designs

set of variants was then evaluated, using weighting factors and an evaluation of each variant on a scale of 0–10 (Figure 66, overleaf). From this, car type 4 emerges as the preferred basic form, and was used as the concept for more detailed design work. (Source: Pighini *et al.*)

*Example 5: Bicycle
splashguard*

The design project for a detachable, rear-wheel splashguard for mountain bicycles, to protect riders from splashes when using the bicycle in town, was introduced in the quality function deployment method (Example 1). Following the QFD analysis, six alternative design concepts were developed for the splashguard, including tyre-cleaning brush attachments, mud flaps of various kinds and a guard attached to saddle and seat post. In order to evaluate these alternative concepts and select the best for development, a method similar to weighted objectives was used.

This variant on the standard method uses a 'datum' design concept, against which all the others are compared. The datum may be chosen from the new alternatives under consideration,

	Sub-function	Function carrier	Task	Principle of evaluation	Weight factor	Type 1	
A	Offers space, support, protection for people and luggage	Body and frame of car	Optimal form for car	Internal space Protective ability	0.12	5	
B	Generates the power for transmission	Motor and trans-mission	Optimal position for motor	Available space Complexity of transmission	0.08	7	8
C	Supports people in a safe and comfortable way	Seats	Optimal disposition for seats	Safety Comfort Possibility of getting 4 seats	0.08	4	5
D	Offers space for luggage	Luggage room	Optimal position and higher capacity of luggage room	Space used for luggage Internal room	0.08	5	4
E₁ E₂	For entry and exit For views outside	Doors Windows	Optimal number, dimensions and position of doors and windows	Facilities for entry, exit and putting in luggage Visibility for driver and passengers	0.08	7	7
F	Changes direction of car driving	Steering system	Optimal disposition of steering system	Position of motor Complexity of steering system	0.08	9	10
G	Aesthetic evaluation				0.08	9	
H	Cost evaluation				0.16	4	
I	Safety evaluation				0.24	4	5
	Total sum					5.48	5.88
	Order of merit					9	7

NOTES: 1 Evaluation marks 0 = unacceptable; 1-3 = still acceptable; 4-6 = fair; 7-9 = good; 10 = very good (optimal solution)
2 Total sum = (mark × weighting factor)
3 *The marks in parentheses are for the car with two seats

Figure 66   Evaluation chart for alternative concepts for the small city-car

Type 2		Type 3		Type 4		Type 5		Type 6
8		5		10		7		3
8	9	8	9	8	10	7	10	9
5	6	5	6	9	10	5	6	5
9	8	3	1	10	9	8	6	2
10	9	6	6	10	9	7	7	8
9	10	9	10	7	8	7	8	10

8		8		7		6		10
5		7		6		5		6
4	5	2	3	*3(5)	*3(6)	5	6	6
6.64	6.96	5.32	5.64	*6.96 (7.44)	*7.12 (7.84)	6.04	6.52	6.28
3	2	10	8	2	1	6	4	5

or an existing design may be used as the datum. In this example, a standard bicycle mudguard with quick-release attachments was chosen as the datum.

The set of design objectives and their weights were determined in the QFD analysis. For each objective, each alternative design concept was then judged as either better (+), worse (−) or the same (S), in comparison to the datum. The decision matrix is shown in Figure 67. Totals for the + signs and − signs are given for each concept at the bottom of the matrix. A

Figure 67
Evaluation chart for
alternative concepts for
the bicycle splashguard

	Wt	I	II	III	IV	V	VI	VII
Easy attach	7	+	+	+	+	+	S	D
Easy detach	4	−	+	+	+	+	S	A
Fast attach	3	+	+	+	+	+	S	T
Fast detach	1	+	+	+	+	+	S	U
Attach when dirty	3	+	+	+	+	S	S	M
Detach when dirty	1	−	+	−	+	S	+	
Not mar	10	+	+	+	+	S	S	
Not catch water	7	−	+	−	S	S	S	
Not rattle	8	−	−	−	−	S	S	
Not wobble	7	−	−	−	S	S	S	
Not bend	4	−	−	−	S	−	S	
Long life	11	−	S	−	S	−	S	
Lightweight	7	+	S	S	−	S	S	
Not release accidentally	10	+	S	S	S	S	S	
Fits most bikes	7	+	S	S	S	S	S	
Streamlined	5	−	S	−	−	+	S	
Total +		8	8	6	7	5	1	0
Total −		8	3	7	3	2	0	0
Overall total		0	5	−1	4	3	1	0
Weighted total		1	17	−15	9	5	1	0

weighted overall total for each concept is calculated by summing the positive and negative weights of the relevant objectives.

In this example, concept 2 emerged as the clear leader. However, it has to be remembered that this is in comparison with the datum, and that direct comparisons between the alternative concepts themselves should also be made. Concept 2 was therefore selected as a new 'datum' and comparisons made with concepts 4 and 5. This check confirmed that concept 2 was the preferred alternative. (Source: Ullman.)

Worked example:
Drive unit

This example of the weighted objectives evaluation method uses estimates of money costs rather than the points-scale system. The example arose in the evaluation of five alternative drive units for a machine that processes synthetic fibre. The objectives used in the evaluation were:

1. Cost
2. Standardization with other units on plant
3. Additional features offered as standard
4. Reliability

A monetary value for 'cost' is, of course, immediately available, but is not so obvious for the other objectives. For the second objective, therefore, estimates were made of costs of additional spares required if a non-standardized unit were chosen. For the third objective it was possible to estimate the cost of incorporating the required additional features, on units for which they were not standard.

Estimating a monetary value for reliability was more difficult. It was decided as a first step to assign subjective relative values to each variant, reflecting the designer's degree of confidence in the unit (based on experience), on a scale from 1 (absolute confidence) to 0 (no confidence). This resulted in Table 3.

The cheapest drive unit with a confidence rating of 1 was V_1 (cost: £355). If problems occurred with units V_2 or V_5, they could

Table 3

	V_1	V_2	V_3	V_4	V_5
Probability of no problem	1	0.8	1	1	0.9
Probability of some problems	0	0.2	0	0	0.1

be replaced with a V_1 unit, giving relative costs (allowing £50 freight costs in addition to the basic costs of a V_1 unit) as shown in Table 4.

Table 4

	V_1	V_2	V_3	V_4	V_5
Reliability cost	0	405 x0.2	0	0	405 x0.1
		─────			─────
		£81			£40.5

The objectives were not weighted in any sophisticated way, but were simply ranked in order of importance and given a weight equivalent to their rank order, as follows:

	Rank order	Weight
Objective 1	3	2
Objective 2	1	4
Objective 3	2	3
Objective 4	4	1

This resulted in the table of costs, Table 5.

Table 5

Variant		Objective: 1 Weight: 2	2 4	3 3	4 1	Weighted total cost
V_1	Cost	335	0	50	0	
	Weighted cost	710	0	150	0	860
V_2	Cost	300	40	50	81	
	Weighted cost	600	160	150	81	991
V_3	Cost	435	0	0	0	
	Weighted cost	870	0	0	0	870
V_4	Cost	385	20	0	0	
	Weighted cost	770	80	0	0	850
V_5	Cost	485	60	25	40.5	
	Weighted cost	970	240	75	40.5	1325.5

The figures for weighted total cost suggest that V_4 is the 'best' alternative, having the lowest cost. However, the designer actually preferred to choose V_3. This suggests that the initial weighting of objectives was incorrect. For V_3 to be preferred implies that objective 1 (the basic cost of the unit) is not assigned a weight of 2, but a weight of 1. That is, the ranking and weighting of objectives should have been:

	Rank order	Weight
Objective 1	4	1
Objective 2	1	4
Objective 3	2	3
Objective 4	3	2

Given these alternative weights, the weighted total costs for the five variants can be calculated as:

V_1	505
V_2	772
V_3	435
V_4	465
V_5	881

Thus, with these revised weightings, V_3 is the lowest cost alternative. This example shows that the process of calculating weighted utility values helps to clarify preferences and priorities that might otherwise remain obscure. When they are out in the open they can at least be discussed and justified, and perhaps revised. (Source: Tebay *et al.*)

10 *Improving Details*

A great deal of design work in practice is concerned not with the creation of radical new design concepts but with the making of modifications to existing product designs. These modifications seek to improve a product—to improve its performance, to reduce its weight, to lower its cost, to enhance its appearance, and so on. All such modifications can usually be classified into one of two types: they are either aimed at increasing its value to the purchaser or at reducing its cost to the producer.

The value of a product to its purchaser is what he or she thinks the product is worth. The cost of a product to its producer is what it costs to design, manufacture and deliver it to the point of sale. A product's selling price normally falls somewhere between its cost to the producer and its value to the purchaser.

Designing is therefore essentially concerned with adding value. When raw materials are converted into a product, value is added over and above the basic costs of the materials and their processing. How much value is added depends on the perceived worth of the product to its purchaser, and that perception is substantially determined by the attributes of the product as provided by the designer.

Of course, values fluctuate, depending on social, cultural, technological and environmental contexts, which change the need for, relevance or usefulness of a product. There are also complex psychological and sociological factors which affect the symbolic or esteem value of a product. But there are also more stable and comprehensible values associated with a product's function, and it is principally these functional values which are of concern to the engineering designer.

The *value engineering* method focuses on functional values, and aims to increase the difference between the cost and value of a product: by lowering cost or adding value, or both. In many cases, the emphasis is simply on reducing costs, and the design effort is concentrated onto the detailed design of components, on their materials, shapes, manufacturing methods and assembly processes. This more limited version of the method is known as *value analysis*. It is usually applied only to the refinement of an existing product, whereas the broader value engineering method is also applicable to new designs or to the substantial redesign of a product. Value analysis particularly requires detailed information on component costs.

Because of the variety and detail of information required in value analysis and value engineering, they are usually conducted as team efforts, involving members from different departments of a company, such as design, costing, marketing, production departments, etc.

The Value Engineering Method

Procedure

List the separate components of the product, and identify the function served by each component

One of the ways in which companies seek to better their rivals' products is to buy an example of the competing product, strip it down to its individual components, and try to learn how their own product might be improved in both design and manufacture. This is one way of learning some of a competitor's secrets without resorting to industrial espionage.

The same sort of technique is at the heart of value engineering and value analysis. The first analytical step in the method is to strip a product down to its separate components—either literally and physically, or by producing an itemized parts list and drawings. However, parts lists and conventional engineering drawings are of limited value in understanding and visualizing the components, the ways in which they fit together in the product overall, and how they are manufactured and assembled. So if an actual product, or a prototype version, is not available for dismantling, then something like exploded diagrams of the product are helpful in showing components in three-dimensional form and in their relative locations or assembly sequences.

The purpose of this first step in the procedure is to develop a thorough familiarity with the product, its components and their assembly. This is particularly important if a team is working on the project, since different team members will have different views of the product, and perhaps only limited understandings of the components and their functions. So it is necessary to go through an exhaustive analysis of the sub-assemblies and individual components, and how they contribute in functional terms to the overall product.

Sometimes it is not at all clear what function a component serves or contributes to! This may be found particularly in products which have had a long life and may have gone through many different versions: some components may simply be redundant items left over from earlier versions. However, it may also be the case that components have been introduced to cope with problems that arose in the use of the product, and so any components which may appear to be redundant should not be dismissed too readily. Sometimes redundancy is even deliberately designed into a product in order to improve its reliability.

The objective of this step in the procedure is to produce a complete list of components, grouped as necessary into sub-assemblies with their identified functions. In value engineering, rather than the more limited value analysis applications of this method, a similar objective applies, even though the ultimate intention might be to develop a completely new product, rather than just to make improvements to an existing product. In this case, the starting point might be an existing product against which it has been decided to compete in the market, or an 'archetypal' or hypothetically typical version of the proposed new product.

Determine the values of the identified functions

Questions of value are, of course, notoriously difficult. They are the stuff of political debate and of subjective argument between individuals. Reaching agreement in a team on the 'value' of particular product functions therefore may not be easy. However, it must be remembered that the value of a product means its value as perceived by its purchaser. So the values of product functions must be those as perceived by customers, rather than by designers or manufacturers. Market research must therefore be the basis of any reliable assessment of the values of functions.

The market prices of different products can sometimes provide indicators of the values that customers ascribe to various functions. For instance, some products exist in a range of different versions, with more functions being incorporated in the products at the higher end of the range. Differences in prices should therefore reflect differences in the perceived values of the additional functions. However, customers are likely to 'perceive' a product as a total entity, rather than as a collection of separate functions, and subjective factors such as appearance are often of more importance than objective functional factors. It is said that the solidity of the 'clunk' made by closing a car door is one of the most important factors influencing a customer's perception of the value of a motorcar.

Considerable efforts have been put into trying to quantify perceived values or benefits, particularly in connection with the 'cost–benefit analysis' method used in planning. For example, in transport planning, some of the benefits of a new road or bridge can be quantified in terms of the time saved by travellers in using the new facility. Attempts are then made to convert all such benefits (and costs) into monetary terms, so that direct comparisons can be made.

Despite the difficulty of assessing values, it is necessary to make the best attempt one can to rationalize and express the perceived values of component functions. It can be pointless to reduce the costs of components if their values are also being reduced, so that the product becomes less desirable (or 'valuable') to prospective purchasers. If quantified and reliable estimates of values cannot be made, then at least simple assessments of high/medium/low value can be attempted.

Determine the costs of the components

Surprising though it may seem, it is not always easy for a company to determine the exact costs of components used in products. The company's accounting methods may not be sufficiently specific for itemized component costs to be identifiable. One of the useful by-products of a value analysis or value engineering exercise, therefore, can be the improvement of costing methods. Team working in the exercise again becomes particularly relevant, because reliable cost information at sufficient detail may only be obtainable by synthesizing information from different departmental specialists.

It is not sufficient to know just the cost of the material in a component, or even its bought-in cost if it is obtained from a supplier. The value analysis team needs to know the cost of the component as an element of the overall product cost—i.e. after it is fully finished and assembled into the product. Therefore, as well as material or bought-in costs there are labour and machine costs to be added for the assembly processes. It is sometimes suggested that factory overhead costs should also be added, but these can be very difficult to assign accurately to individual components, and instead can perhaps be assumed to be spread equally over all components.

It is important not to ignore low-cost components, particularly if they are used in large numbers (e.g. screws or other fasteners). Even a relatively small cost reduction per item can amount to a substantial overall saving when multiplied by the number of components used.

As well as determining the absolute costs of components, their relative or percentage costs in terms of the total product cost should also be calculated. Attention might then be focused on components or sub-assemblies which represent a significant portion of the total cost.

Search for ways of reducing cost without reducing value, or of adding value without adding cost

This fundamental design stage calls for a combination of both critical and creative thinking. The critical thinking is aimed at what the design *is*, and the creative thinking is aimed at what it *might be*. The concept of stripping down a competitor's product to look for ways of improving on it is a useful one to bear in mind at this stage. It is usually easier to be critical of, and to suggest improvements to someone else's design rather than one's own, and it is this kind of creative criticism that is needed at this final stage.

Attempts to reduce costs usually focus on components and on ways of simplifying their design, manufacture or assembly, but the functions performed by a product should also be looked at critically, because it may be possible to simplify them, reduce their range, or even eliminate them altogether if they are of limited value to the purchaser.

There are some general strategies which can be applied in order to direct the search for ways of reducing costs. The first is to concentrate on high-cost components, with a view to substituting lower-cost alternatives. The second is to review any components used in large numbers, since small individual savings may add up to a substantial overall saving. A third strategy is to identify components and functions which are matched as high-cost/high-value, or low-cost/low-value, since the aim is to achieve high-value functions with low-cost components. One particular technique is to compare the cost of a component used in the design with the absolute lowest-cost means of achieving the same function; large differences suggest areas for cost reduction, even though the lowest-cost version may not be a viable option.

A checklist of cost-reduction guidelines can be followed:

Eliminate	Can any function, and therefore its components, be eliminated altogether? Are any components redundant?
Reduce	Can the number of components be reduced? Can several components be combined into one?
Simplify	Is there a simpler alternative? Is there an easier assembly sequence? Is there a simpler shape?
Modify	Is there a satisfactory cheaper material? Can the method of manufacture be improved?
Standardize	Can parts be standards rather than specials? Can dimensions be standardized or modularized? Can components be duplicated?

Whilst the value analysis approach tends to emphasize reducing costs, the broader value engineering approach looks also for ways of adding value to a product. For example, rather than eliminating functions, as suggested above, value engineering might seek ways of improving or enhancing a product's functions. Nevertheless, the aim is always to increase the value/cost ratio.

One of the most significant means of adding value to a product, without necessarily increasing its cost, is to improve its ease of use. This has become particularly evident with the preference for personal computers which are found to be 'user friendly'. In this case, the 'friendliness' perhaps applies more to the computer's software than its hardware, or at least to the combination of software and hardware such that use of the computer seems natural and easy. However, similar principles can be applied to all machines—their use should be straightforward, clear and comfortable. There is a considerable body of knowledge in the field of ergonomics which can be applied to these user aspects of machine design.

Other attributes which commonly contribute to the quality or value of a product are:

Utility	Performance on aspects such as capacity, power, speed, accuracy or versatility.
Reliability	Freedom from breakdown or malfunction; performance under varying environmental conditions.
Safety	Secure, hazard-free operation.
Maintenance	Simple, infrequent or no maintenance requirements.
Lifetime	Except for disposable products, a long lifetime which offers good value for the initial purchase price.
Pollution	Little or no unpleasant or unwanted by-products, including noise and heat.

Finally, there is a whole class of value attributes related to aesthetics. This includes not only the appearance of a product —colour, form, style, etc.—but also aspects such as surface finish and feel to the touch.

Evaluate alternatives and select improvements

The application of value analysis or value engineering should result in a number of alternative suggestions for changes to the product design. Some of these alternatives might well be incompatible with each other, and in fact all suggestions should be carefully evaluated before selecting those which can be shown to be genuine improvements.

Summary

Aim

To increase or maintain the value of a product to its purchaser whilst reducing its cost to its producer.

Procedure

1. List the separate components of the product, and identify the function served by each component.
If possible, the actual product should be disassembled into its components; exploded diagrams and component-function charts are more useful than parts lists.

2. Determine the values of the identified functions.
These must be the values as perceived by customers.

3. Determine the costs of the components.
These must be after fully finished and assembled.

4. Search for ways of reducing cost without reducing value, or of adding value without adding cost.
A creative criticism is necessary, aimed at increasing the value/cost ratio.

5. Evaluate alternatives and select improvements.

Examples

Example 1: Ceiling diffuser

Substantial cost savings can often be made even on relatively simple products. Although the cost-per-unit may not be great, the total savings can be large when a large number of units is involved. In this example, the product is a $10 ceiling diffuser, the device which covers the ceiling output points of heating and air conditioning systems. Its function is to help spread the air flow into the room, and to look attractive.

Figure 68
Value analysis of an air conditioning ceiling diffuser

Informal discussions with a number of users and customers (including installers), to determine their likes and dislikes about

Function	Cost	Design Change
Style product	$1.34 (13%)	Delete the center cone.
Minimize housekeeping	$0.05 (0.5%)	Users complained about ceiling smudging caused by air output. The team reshaped the remaining three cones to feather the air gradually, to prevent it from contacting the ceiling around the diffuser.
Ensure stability	$1.36 (13%)	Team referred to its function-cost worksheet and found that half of the webs, web clips and rivets, all of the springs, and much of the assembly labour was to ensure stability. Changed to two wireforms with two of the legs spot-welded to a newly designed center cone.
Ease installation and simplify adjustment	$1.07 (10%)	Both of these functions are performed by access areas in the cones, to permit the use of a screwdriver to attach the unit to the ductwork and to adjust the ductwork damper after installation. Installation access areas were deleted, since modern installation does not require screw attachment. A hole was added to permit damper adjustment.
Protect shipments	$0.86 (8%)	The prestudy discussions with users/customers revealed that diffusers were invariably ordered in pairs. The shipping carton was redesigned to carry two units at a significant cost reduction.

the existing diffuser, revealed several areas where the designers felt that there were mismatches between manufacturing costs and perceived value to the customers of certain details. The designers' recommendations for changes are given in Figure 68.

The changes resulted in reducing material costs by 24% and labour costs by 84%, saving the company nearly $500 000. Functional and aesthetic improvements also resulted from the redesign. These, plus a 20% price reduction, helped achieve a significant increase in the company's market share. (Source: Fowler.)

Example 2: Ignition system

The product under investigation in this example was a 1200-horsepower engine, used primarily in compressor systems. The engine incorporated several features or subsystems, not provided by competitors. One of these was a totally shielded ignition system, to prevent any possibility of a spark igniting any flammable gas around the engine. The company assumed that this was a valued feature of its engine. However, questionnaires to a panel of users, as part of a value engineering exercise, revealed that the shielded ignition system was not rated highly as a feature by the users.

Figure 69 (overleaf) shows the value analysis report drawn up by the design team. They proposed changing to an unshielded ignition system, with the shielded system offered as a higher-cost option for those customers who wanted it. The estimated cost of making the change was only $1520 for design and drafting. Gross annual saving was $54 000, with a $253 000 net saving over the first three years. (Source: Fowler.)

Example 3: Air valve

This example, the value analysis of an aircraft air valve (Figure 70, overleaf), shows how components and functions can be costed in a comparison table or matrix. Components often contribute to several different or related functions, and hence the cost of a particular function is often spread across several components. The kind of component/function cost matrix shown in Figure 71 allows the designer to analyze in detail these often complex relationships. When a component contributes to more than one function, it may be difficult to break down its overall cost into precise part-costs per function. Approximate but well-informed estimates then have to be made.

The analysis in this example revealed the relatively high cost of the 'connect parts' function, as well as the redundancy of

Figure 69   Value analysis of the ignition system of an engine

Parts or assemblies	Cost (£)
Banjo assembly	1.07
Valve body	6.62
Spring	0.39
Diaphragm assembly	2.14
Cover	2.24
Lug	0.10
Nuts, bolts and washers	2.34
Assembly cost	4.58
Total	£19.48

Figure 70 Cost analysis of an aircraft air valve

(a)

Parts	Stop air	Sense ram air	Sense servo air	Sense cabin air	Connect parts	Provide mounting	Resist corrosion	Provide support	Provide interchangeability	No function	£ total cost	%
Banjo assembly		0.2			0.4				0.47		1.07	5.5
Valve body	0.4	1.0			2.82	0.8	0.2	0.8		0.6	6.62	34.0
Spring										0.39	0.39	2.0
Diaphragm assembly	0.6	0.1	0.1	0.1	0.94		0.2	0.1			2.14	11.0
Cover		0.4			1.2	0.1	0.1	0.34	0.1		2.24	11.5
Lug										0.1	0.1	0.5
Nuts, bolts and washers					2.14	0.1		0.1			2.34	12.0
Assembly					4.58						4.58	23.5
Total	1.0	1.1	0.7	0.1	12.08	0.9	0.6	1.24	0.67	1.09	19.48	100.0
% total	5.1	5.7	3.4	0.5	62.0	4.6	3.1	6.4	3.4	5.6		
High or low					H					H		

(b)

Parts	Stop air	Sense ram air	Sense servo air	Sense cabin air	Connect parts	Provide mounting	Resist corrosion	Provide seal	Provide interchangeability	Provide testing	£ total cost	%
Cover and connection	0.15	0.25	0.50	0.10	0.25	0.30		0.15	0.06		1.76	25.5
Body assembly	0.15	0.20	0.25	0.45	0.45	0.40		0.25	0.03		2.18	31.5
Diaphragm assembly	0.15	0.10	0.25	0.20	0.25	0.10		0.20	0.03		1.28	18.5
Valve assembly	0.05		0.05	0.05	0.15			0.31	0.05		0.66	9.5
Fasteners, nut bolts, etc.					1.04						1.04	15.0
Total	0.50	0.55	1.05	0.80	2.14	0.80		0.91	0.17		6.92	100.0
% total	7.2	7.9	15.1	11.6	30.9	11.6		13.2	2.5			
High or low					H							

Figure 71 (a) Function/cost analysis matrix for the air valve; (b) Function/cost analysis matrix for the redesigned air valve

some elements. A redesign enabled some substantial reductions to be made, with a total cost saving of over 60%. (Source: EITB.)

Example 4: Piston

The elimination of unnecessary parts can be a significant factor in reducing the overall cost of an assembly, and is a principal focus of value analysis. Figure 72 shows the redesign of a small piston assembly, eliminating or combining several parts that were in the original. Separate fasteners should be eliminated wherever possible, and it was found in this example that the two screws could be eliminated by changing the cover plate from steel to plastic with a snap-fit onto the main block. The cover was also redesigned to incorporate the piston stop in a one-piece item. In the redesign, the number of parts was thus almost halved, resulting in reduced material and assembly costs with no loss of performance and an aesthetically improved product. (Source: Redford.)

Figure 72
Redesign of a piston assembly to reduce the number of components

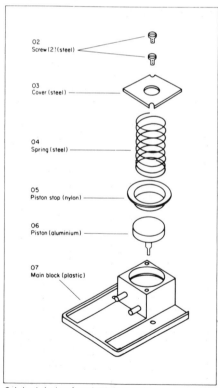

Original design for piston assembly

Redesigned piston assembly

Example 5: Tubular heater

In a company manufacturing various kinds of electrical appliances, its range of tubular heaters was selected for a value engineering exercise. These heaters are simple and robust and used mainly in industrial and office premises to provide background heat. The product range consists of similar tubes of various lengths providing various heat outputs at a standard wattage per unit length.

A component/function/cost analysis, shown in Figure 73, revealed that the largest parts and labour cost was accounted for by what was regarded as the third most important function—that of providing the power connection. A closer examination of this function revealed two distinct sub-functions: firstly, providing an interconnector to allow tubes to be banked together on one mains connection, and secondly, providing a complex terminal connection.

An ideas-generating session produced suggestions for redesign which are shown in Figures 74 and 75 (overleaf). The moulded inter-connector was replaced with three separate wires and a cover piece (Figure 74), which also enabled the terminal itself to be considerably simplified (Figure 75). Together, the modifications resulted in a cost reduction of 21%.

Figure 73
Function/cost analysis of the tubular heater

FUNCTION COST ANALYSIS FORM — Date — Project

Basic function: DISSIPATE HEAT

Item	Support functions	1 ENCLOSE ELEMENT	2 GENERATE HEAT	3 CONNECT POWER	5 PROVIDE APPEARANCE	4 PROTECT AGAINST DAMAGE	Cost of item (P)	%
Tube assy.		24.1					24.1	
Element assy.		6.8	24.8	0.6			32.2	
Term. collar assy.				14.9			14.9	
Cover assy.				4.1			4.1	
Final assy.		0.5	0.5				1.0	
Packing						5.4	5.4	
Painting					4.0		4.0	
Interconnector				13.8			13.8	
Cost of function		31.4	25.3	33.4	4.0	5.4	99.5	
% of total cost		31	25	34	4	5	≈100%	
Cost sequence		②	③	①	⑤	④		

REPORT SUMMARY FORM

Project .....*Heat Tube 5ft.*..................................... Date ....

Item ......*Tube interconnector*............................

Present

Brass earth plate

Copper wires (moulded in)

DMC compression moulded body

Proposed

3 insulated copper wires with spade connectors

Injection moulded shroud

Comments

    This proposal must be phased in at the same time as the termination proposal shown later. Tests have shown the proposal to be easier to install than the present method and there is an improved appearance

Costs		Material	Labour	Additional benefits	Total	
Present		13·0 p	0·8 p		13·8 p	
Proposed		4·0 p	—	Optional item	4·0 p	
Difference	% Saving	9·0p 69%	0·8p 100%		9·8p 71%	

Implementation time ......*4 months*...... Cost £ *1000* ....

Quantity per year ......*70,000*...... Expected life ......*5 yrs.* ......

1st year savings £ ...*~6000*...... Annual savings £ ...*~7000*......

Figure 74
Proposals for redesign of
the tube interconnector

REPORT SUMMARY FORM

Project ....*Heat tube 5ft.*........................ Date ....

Item ....*Terminal cover assembly*.....................

Present

Proposed

Screw to accept brass earth plate or earth wire

Grub screws for interconn. copper wires or supply wires

Bolted-on element Tags

New top cap or shroud

Tag for earth wire

Tags for spade conns. or supply wires.

Push and twist to secure in insulated end plate

21 parts

7 parts in total

Comments

This proposal must be phased in at same time as the tube interconnector proposal shown earlier.

Novel assembly and revised earthing. Tests show no installation or wiring problems. Little change in appearance

Costs	Material	Labour	Additional benefits	Total	
Present	10·6	9·0		19·6	
Proposed	5·3	3·0	Lower stocks etc.	8·3	
Difference / % Saving	5·3p 50%	6·0p 66%		11·3p 58%	

Implementation time ....*6 months*.... Cost £ *1500*

Quantity per year ...*140,000*.... Expected life ..*5 yrs.*

1st year savings £ ...*~14,500*.... Annual savings £ ...*~16,000*

Figure 75
Proposals for redesign of
the terminal cover
assembly

This relatively simple example demonstrates the principles both of applying value analysis with the objective of reducing a product's cost, and of applying value engineering with the objective of generating a more highly-valued, innovative product.

Figure 76 shows how both value analysis and value engineering projects might start—with an exploded diagram of the product, which in this case is a conventional handtorch. The diagram shows the separate components and indicates how they are assembled together in the complete product.

Market research showed that two main aspects of a torch are highly valued by users. These are, firstly, the quality of the emitted light, perceived by users as being influenced by (apart from battery power) the bulb and the reflector, and secondly the ease of use of the torch, determined by the torch body and the switch. One low-valued feature of this particular torch design was the hanging loop on the base of the torch, which was hardly used at all and thus thought by most users to be redundant.

The components, their functions and perceived values are listed in Table 6, with values categorized simply as high, medium or low. It is useful to note that some components which may be important to the technical performance of the product are not necessarily perceived as of high value by users— examples here include the bulbholder and the pressure spring in the base.

A value analysis exercise led fairly quickly to some suggested modifications which would lower the product's cost without lowering its value. The reflector cover seemed to be too complicated, with its three separate components—glass, washer and screw-on retainer. A one-piece, clear plastic cap was suggested as an alternative. The base of the torch also seemed to be a rather complicated assembly, and again a one-piece plastic screw-on cap was suggested, with an integral plastic tongue spring to provide the pressure on the batteries, and the hanging loop eliminated. A proposal was also made to eliminate the switch, with electrical interrupt being provided instead by twisting the head of the torch. However, on evaluation it was decided that this was not very convenient for the user, and risked losing the highly-valued ease of use of the thumb-switch.

Table 6 shows that the costed redesign indicated a potential saving in manufacturing costs of approximately 20%.

A more comprehensive value engineering exercise would have concentrated on the high-value aspects of the torch as

Figure 76
Exploded diagram of a
handtorch

perceived by users, and would have sought to improve these features, to enhance them, or to generate innovations related to them. The high-value features of the torch are to do with its light beam and its handling.

Some research with users might well have found that the conventional torch has some shortcomings in these areas. For instance, it seems basically designed to throw a moderately

Table 6

Component	Function	Value	Cost(£) Original	Cost(£) Redesign
Cap Washer Glass	Protect bulb and reflector	Medium	0.16	0.08
Reflector	Project light beam	High	0.12	0.12
Bulb	Provide light	High	0.10	0.10
Bulb holder	Hold bulb, provide electrical cointact	Low	0.05	0.05
Torch body	Contain batteries, locate parts, provide hand grip	High	0.26	0.26
Switch	Provide electrical interrupt	High	0.08	0.08
Spring Washer	Provide pressure on batteries	Low	0.10	0.10
Cap	Protect batteries	Medium	0.10	
Loop	Provide for hanging	Low	0.03	—
Total			£1.00	£0.79

Figure 77
The Durabeam handtorch

wide beam over a fairly large distance—such as for illuminating a footpath. But most use of a torch these days is for closer illumination such as finding a keyhole or making emergency repairs on a motorcar engine. In the latter type of case, it is important to be able to place the torch down, leaving one's hands free, and to direct the beam to the appropriate spot. The conventional, cylindrical torch is poorly designed for this; it is also inconveniently shaped for carrying in a pocket or handbag.

The novel 'Durabeam' torch (Figure 77) illustrates how these principles might have been applied in the design of a new product. The batteries are placed side-by-side instead of end-to-end, creating a flat, rectangular, compact body shape. The thumb-switch has been eliminated by using a 'flip-top' mechanism which acts as a switch and which also allows the angle of direction of the beam to be adjusted.

11 *Design Strategies*

Using any particular design method during the design process
will often appear to be diverting effort from the central task of
designing. However, this is exactly the importance of using
such a method—it involves applying some thought to the *way*
in which the problem is being tackled. It requires some strategic
thinking about the design process.

A design strategy describes the general plan of action for a
design project and the sequence of particular activities (i.e. the
tactics, or design methods) which the designer or design team
expect to take to carry through the plan. To have a strategy is
to be aware of where you are going and how you intend to get
there. The purpose of having a strategy is to ensure that activities remain realistic with respect to the constraints of time,
resources, etc. within which the design team has to work.

Many designers seem to operate with no explicit design
strategy. However, having no apparent plan of action can be a
strategy, of sorts! It might be called a 'random search' strategy,
and might very well be appropriate in novel design situations
of great uncertainty, where the widest possible search for solutions is being made. Examples of such novel situations might be
trying to find applications for a completely new material, or
designing a completely new machine such as a domestic robot.

For these kinds of situations, an appropriate strategy would
be to search (at least to begin with) as widely as possible, hoping
to find or generate some really novel and good ideas. The
relevant tactics would be drawn mainly from the creative
methods.

At the opposite extreme to 'random search' would be a
completely predictable or 'prefabricated' sequence of tried-and-
tested actions. Such a strategy would be appropriate in familiar
and well-known situations. Again, it might not seem to be an
explicit strategy, simply because it involves following a well-

worn path of conventional activities. Examples of appropriate situations for such a strategy might include designing another variation of the machine that the designer's employer always makes, or designing a specific and conventional type of product for an identified sector of the market.

In such situations, the design strategy would be aimed at narrowing the search for solutions and quickly homing-in on a satisfactory design. Relevant tactics would be drawn from conventional techniques and the rational methods.

Strategy Styles

The 'random search' and 'prefabricated' strategies represent two extreme forms. In practice, most design projects require a strategy that lies somewhere between the two extremes, and contains elements of both.

The 'random search' strategy represents a predominantly *divergent* design approach; the 'prefabricated' strategy represents a predominantly *convergent* approach. Normally, the overall aim of a design strategy will be to converge onto a final, evaluated and detailed design proposal. But within the process of reaching that final design there will be times when it will be appropriate and necessary to diverge—to widen the search or to seek new ideas and starting points.

The overall design process is therefore convergent, but it will contain periods of deliberate divergence (Figure 78).

Psychologists have suggested that some people are more naturally convergent thinkers and some are more naturally divergent thinkers. These preferred 'thinking styles' mean that some designers may be happier with one kind of strategy style rather than with another; one person may prefer a more convergent style, whereas another may prefer a more divergent style.

Figure 78
The overall design process is convergent, but it includes periods of both convergence and divergence

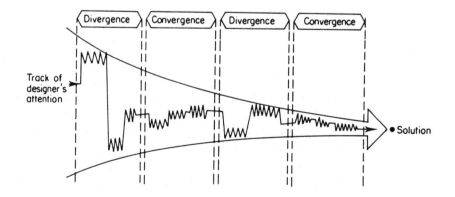

Alternatively, in a team context, designers with one preferred style may come to the fore in certain stages of the design process, and others come to the fore at other stages.

Convergent thinkers are usually good at detail design, at evaluation and at selecting the most appropriate or feasible proposal from a range of options. Divergent thinkers are usually good at concept design and at the generation of a wide range of alternatives. Clearly both kinds of thinking are necessary for successful design. Unfortunately, much engineering (and other) education tends to promote and develop only convergent thinking.

As well as convergent and divergent, other kinds of 'thinking styles' have also been identified by psychologists, and may also have importance in design and in the structuring of design strategies. One of the most important dichotomies in thinking style appears to be that between *serialist* and *holist*. A serialist thinker prefers to proceed in small, logical steps, tries to get every point clear or decision made before moving on to the next, and pursues a straight path through the task, trying to avoid any digressions. A holistic thinker prefers to proceed on a much broader front, picking up and using bits of information that are not necessarily connected logically, and often doing things 'out of sequence'.

Another distinction that has been made between styles of thinking is that between *linear* and *lateral* thinking. Linear thinking proceeds quickly and efficiently towards a perceived goal, but may result in getting 'stuck in a rut', whilst lateral thinking entails a readiness to see, and to move to, new directions of thought.

The dichotomies of thinking style suggested by the psychologists tend to fall into two groups:

Convergent	Divergent
Serialist	Holist
Linear	Lateral

There is even some evidence to suggest that there is a fundamental dichotomy between the 'thinking styles' of the two hemispheres of the human brain. The left hemisphere predominates in rational, verbal, analytic modes of thought, whilst the right hemisphere predominates in intuitive, nonverbal, synthetic modes of thought.

Differences of thinking styles therefore appear to be an inherent characteristic of human beings. Most people *tend* towards a preference for one style rather than another; but no-one is *exclusively* limited to just one style. In particular, it is actually important to be able to change from one style to another in the course of a design project.

However, many models of the design process, such as those discussed in Chapter 2, do tend to present design as a linear, serialistic process. This may be off-putting, and even counter-productive to those designers whose own preferred thinking style tends more towards the lateral and holistic. What is needed is a more flexible, strategic approach to designing, which identifies and fosters the right kind of thinking at the right time, and within the context of the particular design project.

Strategy Analogies

To convey this more flexible approach to design strategies and tactics, some authors have resorted to the use of analogies. For example, Jones has suggested that a designer is like an explorer searching for buried treasure:

> A new problem is like an unknown land, of unknown extent, in which the explorer searches by making a network of journeys. He has to invent this network, either before he starts or as he proceeds.
>
> Design methods are like navigational tools, used to plot the course of a journey and maintain control over where he goes. Designing, like navigation, would be straightforward if one did not have to depend on inadequate information in the first place. Unlike the explorer's, the designer's landscape is unstable and imaginary, it changes form according to the assumptions he makes.
>
> The designer has to make as much sense as he can of every fragmentary clue, so that he can arrive at the treasure without spending a lifetime on the search. Unless he is very unlucky, or very stupid, he will come across the treasure long before he has searched every inch of the ground.

Koberg and Bagnall have suggested that the designer is like a traveller, and that 'the design process is a problem-solving journey':

> A general rule is to find and use those methods which best fit the problem as well as the abilities of the problem solver. It's a task similar to that of selecting the route, side roads and overnight stops for an auto trip. Just as any competent trip planner would examine the alternative routes on a map, and read through several

brochures, books or articles before choosing a route for his trip, so should the problem solver review the methods available, and not be afraid to adapt any of them to his special needs.

Instead of exploring or travelling, I prefer an analogy based on football. A design team, like a football team, has to have a strategy. The football team's strategy for defeating the opposition will consist of an agreed plan to use a variety of plays or moves (i.e. techniques or methods), to be applied as the situation demands. During the game, the choice of a move, and whether or not it is successful, will depend on the specific circumstances, on the skill of the players, and on the response of the opposition.

The repertoire of moves used in a game is partly decided in advance, partly improvised on the field, and also amended at the half-time briefing by the team coach. The coach's role is important because he maintains a wider view of the game than the players can actually out there on the field. In designing, it is necessary to adopt a similar role from time to time, in reviewing the project's strategy and progress.

For you as an individual designer, or member of a design team, tackling your problem and reaching your goal will involve both the strategic skills of the coach and the tactical skills of the player. Also, like the team, you will have to make on-field and half-time reviews of your strategy to ensure that your problem doesn't defeat you! A design strategy, therefore, should provide you with two things:

1. A *framework* of intended actions within which to operate,
2. A management *control* function enabling you to adapt your actions as you learn more about the problem and its 'responses' to your actions.

Frameworks for Action

One 'framework', complete with appropriate methods identified and located within it, has already been suggested. That was the procedural model of the design process that I outlined in Chapter 3:

Stages in the design process	Appropriate methods
1. Clarifying objectives	Objectives tree
2. Establishing functions	Function analysis
3. Setting requirements	Performance specification
4. Determining characteristics	Quality function deployment

5.   Generating alternatives   Morphological chart
6.   Evaluating alternatives   Weighted objectives
7.   Improving details   Value engineering

If it seemed to be appropriate to the specific project in hand, then you could adopt this as a complete 'prefabricated' strategy. It comprises a seven-stage framework covering the design process from client objectives through to detail design, and a suitable 'tactic'—a design method—for each stage. You could, of course, add, or substitute, methods in each stage. For example, you could use brainstorming instead of a morphological chart as a way of generating alternative solutions; you could use the conventional 'design-by-drawing' method instead of, or perhaps as well as, value engineering or analysis at the stage of detail design.

However, this particular framework does imply that the design process is going to be a fairly straightforward, step-by-step process. It implies a 'linear' design process. A design strategy that is more suited to a 'lateral' approach might be something like this:

	Stage	*Methods to be used*
1.	Divergent problem exploration	Morphological Chart
		Brainstorming
2.	Structuring of problem	Objectives Tree
		Performance Specification
3.	Convergence on solution	Synectics

Another framework might be adopted from the general pattern of the creative process, as outlined in Chapter 3. This might be developed as follows:

	Stage	*Methods to be used*
1.	Recognition	Brainstorming
		Writing a design brief
2.	Preparation	Objectives tree
		Information search
		Function analysis
3.	Incubation	Taking a holiday
		Talking the problem over with colleagues and friends
		Tackling another problem
		Enlarging the search space: counterplanning

4.	Illumination	Morphological chart
		Brainstorming
		Enlarging the search space:
		random input
5.	Verification	Performance specification
		Weighted objectives

So you see that there can be many different strategy frameworks, and many different tactical combinations of methods and techniques.

Strategy Control

The second important aspect of a successful design strategy is that it has a strong element of management control built into it. If you are working alone on a project, then this means self-management, of course. If you are working in a team then either the team leader or the whole team, collectively, must from time to time review progress and amend the strategy and tactics if necessary.

Whatever general framework is adopted for the project, it is necessary to have some further strategy control in order to avoid unnecessary time-wasting, going down blind alleys, and the like. Some simple rules of strategy control are:

1. Keep your objectives clear
In designing, it is impossible to have one set of completely-fixed objectives, because ends and means are inextricably interwoven in the product you are designing. A creative resolution of a design problem often involves changing some of the original objectives. However, this does not mean that it is impossible to have any clear objectives at all. On the contrary, it is important to have your objectives clear at any time (probably in the form of an 'objectives tree'), but also to recognize that they can change as your project evolves.

2. Keep your strategy under review
Remember that your overall aim is to solve the design problem in a creative and appropriate way—it is not doggedly to follow a path you have set for yourself that might be leading nowhere! A design strategy needs to be flexible, adaptable and intelligent; so review it regularly. If you feel that your actions are not being very productive, or that you're getting stuck, then pause to ask yourself if there is a better way of proceeding. Have confidence in adapting the 'tactics'—the methods and techniques—to

your own ways of working and to the aims and progress of the project.

3. *Involve other people*
Different people 'see' a problem in different ways, and it is often true that 'two heads are better than one'. If you are getting stuck, one of the easiest ways to sort out what is going wrong is to explain the project to someone else—a colleague or a friend. Other people, of course, are also able to offer ideas and different viewpoints on the problem which may well suggest ways to change your approach.

4. *Keep separate files for different aspects*
There will, almost certainly, be times when you are having to work on several different aspects of a project in parallel. So keep separate files which allow you to switch rapidly from one aspect to another, or to take in a new piece of information in one area without distracting your work on another. One very useful file to keep is 'Solution Ideas'. You will probably come across, or have ideas for solutions at all times throughout the project, but you will need just to keep them filed until you are ready to turn your whole attention to solution concepts or details.

Exercises: Choosing Strategies and Tactics

These short exercises are intended to give some practice in devising strategic 'frameworks' and selecting appropriate tactical methods or techniques. Each exercise need only take 5–10 minutes.

Exercise 1

Your company manufactures industrial doors of various kinds. With the increased availability of electronic devices, remote controls, and so on, the company has decided to produce a new range of automatically-operated doors. You have been asked to propose a set of prototype designs that will establish the basic features of this new range. Outline your design strategy and tactics.

Exercise 2

Your company manufactures packing machinery. One of the company's most valued customers is about to change its product range and will therefore need to replace its packing machinery. You will be responsible for designing this new machinery. Outline your design strategy and tactics.

Exercise 3

You have just been appointed design consultant to a company manufacturing office equipment. Its sales have fallen drastically because its designs have failed to keep up with modern office equipment trends. To re-establish its position the company wants a completely new product that will be a step ahead of all its rivals. You have to suggest what the new product should be, and produce some preliminary design proposals for a Board meeting in two weeks' time. Outline your design strategy and tactics.

Discussion of Exercises

1. Industrial doors

The change from manual to automatic doors implies that there could be scope for rethinking the extent of the company's current range—perhaps to include some door types that were not previously included. It is therefore worth putting some divergent search effort into the early stages of the design project. It is also important not to overlook the features of existing doors that are valued by customers, so quality function deployment could be used to identify the critical characteristics. My suggested strategy would be:

	Framework	*Methods*
1.	Problem exploration	Brainstorming
		Synectics
2.	Problem specification	Quality function deployment
3.	Alternative solutions	Morphological chart
4.	Selection of alternatives	Weighted objectives

2. Packing machinery

This appears to be a straightforward case of redesigning an established product. There is no apparent need for radically new design concepts, so fairly conventional methods can be used. My suggested strategy would be:

	Framework	*Methods*
1.	Customer requirements/ Problem specification	Performance specification
2.	Alternative solutions/ Evaluation of alternatives	Value engineering
3.	Detail design	Conventional design-by-drawing

3. Office equipment

This problem does suggest the need for some radical design thinking, and pretty quickly! Creativity techniques would therefore feature strongly in the strategy. Following the generation of ideas, I think that I might try to use a modified version of the objectives tree method in a final, more convergent stage, and to work this into my presentation to the Board so as to relate the choice of alternatives to the company's objectives. My suggested strategy would be:

	Framework	*Methods*
1.	Divergent search	Enlarging the search space: why–why–why? Function analysis
2.	Alternative solutions	Brainstorming Morphological chart
3.	Convergent selection	Objectives tree

I hope that these brief examples give an indication of how to adopt a strategic approach to product design, using a variety of methods as the tactics of designing. The important points to remember are to devise a strategy that responds to the particular problem and situation, to keep your strategy flexible and to review its effectiveness from time to time during the design project.

References and Sources

Archer, L. B. (1984) *Systematic Method for Designers*. In: N. Cross (ed.), *Developments in Design Methodology* Wiley, Chichester.

Davies, R. (1985) A psychological enquiry into the origination and implementation of ideas *MSc Thesis* Dept. of Management Sciences, University of Manchester Institute of Science and Technology.

Ehrlenspiel, K. and John, T. (1987) Inventing by design methodology *International Conference on Engineering Design*, American Society of Mechanical Engineers, New York.

Engineering Industry Training Board (no date) *Value Engineering* E. I. T. B., London.

Fowler, T. C. (1990) *Value Analysis in Design* Van Nostrand Reinhold, New York.

French, M. J. (1985) *Conceptual Design for Engineers* Design Council, London.

Hauser, J. R. and Clausing, D. (1988) The house of quality *Harvard Business Review* May-June.

Hawkes, B. and Abinett, R. (1984) *The Engineering Design Process* Pitman, London.

Hubka, V. (1982) *Principles of Engineering Design* Butterworth, London.

Hubka, V., Andreasen, M. M. and Eder, W. E. (1988) *Practical Studies in Systematic Design* Butterworths, London.

Jones, J. C. (1981) *Design Methods* Wiley, Chichester.

Jones, J. C. (1984) *A Method of Systematic Design*. In: N. Cross (ed.), *Developments in Design Methodology* Wiley, Chichester.

Koberg, D. and Bagnall, J. (1974) *The Universal Traveler* Kaufmann, Los Altos, Calif.

Krick, E. (1976) *An Introduction to Engineering* Wiley, New York.

Lawson, B. R. (1984) *Cognitive Strategies in Architectural Design*. In: N. Cross (ed.), *Developments in Design Methodology* Wiley, Chichester.

Love, S. F. (1980) *Planning and Creating Successful Engineered Designs* Advanced Professional Development, Los Angeles, Calif.

Luckman, J. (1984) *An Approach to the Management of Design*. In: N. Cross (ed.), *Developments in Design Methodology* Wiley, Chichester.

March, L. J. (1984) *The Logic of Design*. In: N. Cross (ed.), *Developments in Design Methodology* Wiley, Chichester.

Marples, D. (1960) *The Decisions of Engineering Design* Institute of Engineering Designers, London.

Norris, K. W. (1963) *The Morphological Approach to Engineering Design.* In: J. C. Jones and D. Thornley (eds), *Conference on Design Methods* Pergamon, Oxford.

Pahl, G. and Beitz, W. (1984) *Engineering Design* Design Council, London.

Pighini, U., *et al.* (1983) The determination of optimal dimensions for a city car *Design Studies* **4**, No. 4.

Pugh, S. (1991) *Total Design* Addison Wesley, Wokingham.

Ramaswamy R. and Ulrich, K. (1992) *Augmenting the House of Quality with Engineering Models.* In: D. L. Taylor and L. A. Stauffer (eds), *Design Theory and Methodology* American Society of Mechanical Engineers, New York.

Redford, A. H. (1983) Design for assembly *Design Studies* **4**, No. 3.

Shahin, M. M. A. (1988) Application of a systematic design methodology *Design Studies*, **9**, No. 4.

Simon, H. A. (1984) *The Structure of Ill-Structured Problems.* In: N. Cross (ed.), *Developments in Design Methodology* Wiley, Chichester.

Tebay, R., Atherton, J. and Wearne, S. H. (1984) Mechanical engineering design decisions *Proc. Inst. Mech. Engnrs.* **198B**, No. 6.

Tjalve, E. (1979) *A Short Course in Industrial Design* Newnes-Butterworths, London.

Tovey, M. (1986) Thinking styles and modelling systems *Design Studies* **7**, No. 1.

Ullman, D. G. (1992) *The Mechanical Design Process* McGraw-Hill, New York.

Index